Frederike Neise

Risk Management in Stochastic Integer Programming

VIEWEG+TEUBNER RESEARCH

Frederike Neise

Risk Management in Stochastic Integer Programming

With Application to Dispersed Power Generation

VIEWEG+TEUBNER RESEARCH

Bibliographic information published by the Deutsche Nationalbibliothek
The Deutsche Nationalbibliothek lists this publication in the Deutsche Nationalbibliografie;
detailed bibliographic data are available in the Internet at http://dnb.d-nb.de.

Beim vorliegenden Buch handelt es sich um eine vom Fachbereich Mathematik der Universität Duisburg-Essen genehmigte Dissertation.

Datum der mündlichen Prüfung: 11. Februar 2008

Referent: Prof. Dr. Rüdiger Schultz
Korreferenten: PD Dr. René Henrion, Prof. Dr.-Ing. Edmund Handschin

1st Edition 2008

All rights reserved
© Vieweg+Teubner | GWV Fachverlage GmbH, Wiesbaden 2008

Readers: Christel A. Roß

Vieweg+Teubner is part of the specialist publishing group Springer Science+Business Media
www.viewegteubner.de

No part of this publication may be reproduced, stored in a retrieval system or transmitted, in any form or by any means, electronic, mechanical, photocopying, recording, or otherwise, without the prior written permission of the copyright holder.

Registered and/or industrial names, trade names, trade descriptions etc. cited in this publication are part of the law for trade-mark protection and may not be used free in any form or by any means even if this is not specifically marked.

Cover design: KünkelLopka Medienentwicklung, Heidelberg
Printing company: Strauss Offsetdruck, Mörlenbach
Printed on acid-free paper

ISBN 978-3-8348-0547-8

Preface

I am deeply grateful to my advisor Prof. Dr. Rüdiger Schultz for his untiring encouragement. Moreover, I would like to express my gratitude to Prof. Dr.-Ing. Edmund Handschin and Dr.-Ing. Hendrik Neumann from the University of Dortmund for inspiration and support. I would like to thank PD Dr. René Henrion from the Weierstrass Institute for Applied Analysis and Stochastics in Berlin for reviewing this thesis. Cordial thanks to my colleagues at the University of Duisburg-Essen for motivating and fruitful discussions as well as a pleasurable cooperation.

Contents

1 Introduction

A decision maker often has to choose an action out of alternatives without having complete information about the data defining the problem he is faced with and determining his decision's consequences. Though substantially influenced by uncertainty, his decision should be optimal in some sense with respect to the possible future scenarios. If we assume the uncertain data to be given in form of random variables, such a situation can be described by a stochastic optimization problem.

Stochastic optimization has been pioneered in the 1950's/60's by Dantzig, Beale, Charnes and Cooper. G. B. Dantzig was the first to formulate a linear programming problem including uncertain data, see [Dan55]. E. M. L. Beale ([Bea55]) addressed two-stage stochastic programs with fixed recourse. Whereas A. Charnes and W. W. Cooper introduced a particular stochastic problem with chance constraints in [CC59]. Further early contributions to stochastic programming are [Bea61, DM61, Tin55, vdPP63], for instance.

Stochastic programming has found many applications, for example in airline yield management, capacity planning, unit commitment, financial planning, portfolio optimization, logistics and, telecommunications network planning. We are going to present and solve a problem from energy optimization: The optimal operation of a dispersed generation system with respect to uncertain power and heat demand, gas and power prices, and infeed from renewable resources.

There are many different approaches known in the field of stochastic optimization to formulate, handle, and efficiently solve random optimization problems. In this thesis, we consider two concepts: Mean-risk models involving different risk measures as well as stochastic optimization problems with first order dominance constraints induced by mixed-integer linear recourse. They are both based on classical mixed-integer linear two-stage stochastic optimization which is characterized by an additional information constraint: It is assumed that an initial decision which does not foresee or influence the future is followed by a recourse action that is allowed to observe the initial decision and the realization of the stochastic data.

Both considered concepts are suitable to manage risks in stochastic mixed-integer linear programming. They provide solutions according to the preferences of a decision maker who acts rational and tries to avoid risky actions. Mean-risk optimization aims at finding a "best" decision applying statistical parameters to the stochastic objective function like Mean, Variance, or other moments of higher order. In contrast, dominance constrained programming rather heads towards a decision which dominates some given cost benchmark and simultaneously is optimal according to an additional objective.

Since the more "traditional" mean-risk models have already been analyzed in detail, we merely review some results on structure and stability of these problems and describe algorithmic issues. Moreover, we establish a particular mean-risk formulation for the dispersed generation problem. Using a tailored dual decomposition algorithm, we efficiently solve arising deterministic equivalents. The results show the superiority of a decomposition over standard mixed-integer linear solvers and prove the benefits of stochastic optimization to the management of dispersed generation systems.

Dominance Constraints so far have only been considered for problems involving random variables that enjoy suitable continuity and smoothness properties. The latter is not maintained for dominance constraints induced by mixed-integer linear recourse. We address this new class of problems, where first order dominance is applied to a two-stage stochastic optimization problem with integer requirements in the second stage. We analyze the problem's structure, show its well-posedness, and prove similar stability results as known for mean-risk models, concerning perturbations of the underlying probability measures. Furthermore, we establish deterministic equivalents and present solution procedures. They are based on different relaxations providing lower bounds and heuristics, which generate feasible solutions, embedded into a branch-and-bound scheme. The lower bounding procedures exploit the equivalents' block-structure and allow for dual decomposition according to data scenarios. Finally, first order dominance constrained optimization is applied to the optimal management of dispersed generation, and computations which aim at controlling operational costs and minimizing the units' abrasion are reported.

Some results concerning the application of the mean-risk approach to dispersed power generation have already been published in [HNNS06, SN07]. Moreover, in [GNS06, GGNS07] dominance constraints induced by mixed-integer linear recourse are addressed and their application to dispersed generation is described.

This thesis arises out of a project funded by the German Federal Ministry of Education and Research (BMBF) under grant 03SCNIVG. The project "Verbundoptimierung von dezentralen Energieumwandlungsanlagen" is a cooperation with the Chair of Energy Systems and Energy Economics of the University of Dortmund, especially Prof. Dr.-Ing. E. Handschin and Dr.-Ing. H. Neumann. Related aspects considering mixed-integer linear two-stage stochastic optimization and its application to the optimal operation of a dispersed generation system have been established in [Neu07].

1.1 Stochastic Optimization

We consider a situation where a decision maker has to select an optimal action without having complete information about the data involved. Additionally, we assume that following an initial decision we allow a so-called recourse decision. It anticipates the future, which means that this recourse action is made after the realization of the stochastic data has been observed. This leads to the field of two-stage stochastic optimization. Below, we first introduce a random optimization problem with uncertain right hand side and then describe the deduction of a two-stage stochastic optimization problem regarding the additional information constraint. We end up with a very common concretization of a two-stage program – the expectation-based optimization problem. Note that two-stage stochastic optimization is just one possible extension of a general stochastic optimization problem. There are many other approaches to handle stochastic optimization problems (for examples see [BL97, KM05, KW94]) that shall not be considered here.

1.1.1 The two-stage stochastic optimization problem

We start out from a *random mixed-integer linear optimization problem* with uncertain data in the right hand side

$$\min\left\{ c^\top x + q^\top y : Tx + Wy = z(\omega), \, x \in X, \, y \in \mathbb{Z}_+^{\bar{m}} \times \mathbb{R}_+^{m'} \right\}. \quad (1.1)$$

It aims at minimizing the costs $c^\top x + q^\top y$ induced by decisions x and y, respectively, with cost vectors $c \in \mathbb{R}^m$ and $q \in \mathbb{R}^{\bar{m}+m'}$. $X \subseteq \mathbb{R}^m$ is supposed to be a nonempty polyhedron possibly involving integer requirements to components of x. Moreover, integer requirements to components of y are assumed. $Tx + Wy = z(\omega)$, including conformable matrices $T \in \mathbb{R}^{s \times m}$ and $W \in \mathbb{R}^{s \times \bar{m}+m'}$, denotes all constraints including stochastic data represented by the random variable $z(\omega)$, which lives on an appropriate probability space $(\Omega, \mathscr{A}, \mathbb{P})$ and maps elements of Ω to $\mathbb{R}^s$. Note that c, q, T, and W could be influenced by stochasticity, too. For convenience, we restrict our considerations to random problems with stochastic right hand side in this thesis

With the implicit information constraint that x has to be chosen without anticipation of the realizations of $z(\omega)$, whereas y is allowed to observe the future and the selection of x, a two-stage dynamics of decision and observation enters the problem. We first choose x, then we are allowed to observe $z(\omega)$, before we finally make the decision $y = y(x, \omega)$, which depends on x and ω. This alternating scheme of decision and observation together with the nonanticipativity of x becomes explicit using the following equivalent reformulation of (1.1) as *two-stage stochastic optimization problem*

$$\min_{x} \left\{ c^\top x + \min_{y} \left\{ q^\top y \ : \ Wy = z(\omega) - Tx, \ y \in \mathbb{Z}_+^{\bar{m}} \times \mathbb{R}_+^{m'} \right\} \ : \ x \in X \right\}. \quad (1.2)$$

According to the underlying information constraint, x is called first-stage and y second-stage decision. Defining the *value function of the second-stage problem* by

$$\Phi(t) := \min \left\{ q^\top y \ : \ Wy = t, \ y \in \mathbb{Z}_+^{\bar{m}} \times \mathbb{R}_+^{m'} \right\}, \quad (1.3)$$

leads to the following reformulation of (1.2) which even better reflects the two-stage structure of the problem

$$\min_{x} \left\{ c^\top x + \Phi(z(\omega) - Tx) \ : \ x \in X \right\}. \quad (1.4)$$

Obviously, the chief difficulty is to find an optimal first-stage decision x. The optimal recourse decision y then results from the solution of Φ at the point $z(\omega) - Tx$.

The value function Φ has been studied in parametric optimization. With the additional assumptions

$$W \in \mathbb{Q}^{s \times \bar{m} + m'} \ (W \text{ is rational}), \quad (1.5)$$

$$W(\mathbb{Z}_+^{\bar{m}} \times \mathbb{R}_+^{m'}) = \mathbb{R}^s \ (\text{complete recourse}), \text{ and} \quad (1.6)$$

$$\left\{ u \in \mathbb{R}^s \ : \ W^\top u \le q \right\} \ne \emptyset \ (\text{sufficiently expensive recourse}), \quad (1.7)$$

Φ is real-valued and lower semicontinuous on $\mathbb{R}^s$, which means that

$$\liminf_{t_n \to t} \Phi(t_n) \ge \Phi(t)$$

for all $t \in \mathbb{R}^s$. For details see [BM88, BJ77].

Considering (1.4) and the dependence of Φ on $z(\omega)$, every first-stage decision $x \in X$ obviously induces a real-valued random variable

$$f(x,z) := c^\top x + \Phi(z(\omega) - Tx). \quad (1.8)$$

Thus, finding an optimal first-stage solution x to the two-stage stochastic optimization problem (1.4) corresponds to the selection of a "best" random variable from the family

$$\left(c^\top x + \Phi(z(\omega) - Tx) \right)_{x \in X}. \quad (1.9)$$

Each member $f(x,z)$ of this family represents the stochastic costs induced by a decision x and the resulting optimal recourse action y.

For an overview on stochastic programming we refer to standard books like [BL97, KW94, Pré95, RS03].

1.1.2 Expectation-based formulation

To solve (1.4) we have to select a "best" random variable from a family of random variables. But what makes a random variable being the "best"? Or in other words, when is a random variable preferable to another one? To answer this question, an appropriate ranking of random variables in a minimization context is needed. Several characteristics of random variables come into mind that could be useful to compare them and define a "best" one. Commonly, the mean of random variables is applied which yields the traditional *expectation-based stochastic optimization problem with mixed-integer linear recourse*

$$\min\left\{\mathbb{Q}_{\mathbb{E}}(x) \,:\, x \in X\right\}, \tag{1.10}$$

where

$$\mathbb{Q}_{\mathbb{E}}(x) := \mathbb{E}\left(c^\top x + \Phi(z - Tx)\right) = \int_{\mathbb{R}^s} \left(c^\top x + \Phi(z - Tx)\right) \mu(dz)$$

with

$$\mu := \mathbb{P} \circ z^{-1}$$

the probability measure induced by the random variable $z(\omega)$.

Using the expectation-based model corresponds to a minimization of the expected costs induced by the nonanticipative decision x and the recourse decision y. This means that the random variable $f(x, z)$ is chosen that causes minimum costs in the mean. This is a straightforward way to compare random variables, which, of course, is often used in everyday life to rank alternatives. But it forgets about the variability of a random variable. Hence, importance is attached to future scenarios which are assumed to occur quite probably, whereas scenarios which will occur rarely, but can be very expensive, are neglected. On the one hand this is reasonable if the underlying decision can be repeated an infinite number of times. Then the expectation meets the requirements of a rational decision maker. On the other hand in practice decisions usually are not repeated infinitely. Therefore, if a decision can be made only once, the use of the expectation-based model can lead to unaffordable costs if one of the neglected but very bad scenarios occurs. To overcome this limited view, which forgets about worst cases and riskiness of a decision, we subsequently discuss two other approaches to rank and select one of the random variables induced by different first-stage decisions x: Mean-risk optimization models and first order dominance constrained stochastic optimization problems.

1.2 Content and Structure

This work is organized as follows: The second chapter concerns mean-risk optimization problems. It starts with a definition of different risk measures – deviation measures and measures based on quantiles of probability distributions. We continue with an introduction to mean-risk models and review structural and stability results with respect to perturbations of the underlying probability measure. These results allow for a discrete approximation of the probability distribution and therewith the derivation of deterministic equivalents for mean-risk models. The chapter is concluded with the presentation of a dual decomposition algorithm which is based on Lagrangian relaxation embedded into a branch-and-bound scheme. The algorithm exploits the deterministic equivalents' structure and therefore handles them much more efficiently than standard mixed-integer linear solvers can do.

In the third chapter the theory of stochastic dominance and resulting mixed-integer linear dominance constrained optimization problems are addressed. We first introduce stochastic dominance relations, which form partial orders among random variables. First and second order stochastic dominance are defined for the preference of higher outcomes as well as for the preference of smaller outcomes of a decision maker (maximizing profit/minimizing losses). Moreover, some equivalent definitions of stochastic dominance and relations between first and second order dominance are given. We continue with the detailed description of a stochastic optimization problem with first order dominance constraints induced by mixed-integer linear recourse. We prove the well-posedness of this problem and establish results on structure and stability with respect to perturbations of the underlying probability measure. Again, these results are a justification for the derivation of an equivalent, deterministic formulation, which arises when the involved probability distributions are assumed to be given by finitely many scenarios and probabilities. Furthermore, the chapter deals with algorithmic aspects considering the solution of the deterministic equivalent. We propose different approaches to compute lower bounds exploiting the problem's structure and allowing for decomposition of the large-scale problems, discuss their advantages and disadvantages, and develop a heuristic to find feasible solutions. The chapter concludes with a branch-and-bound algorithm.

The fourth chapter is concerned with the optimal operation of a dispersed generation system with respect to the supply of heat and power demand, where the above described optimization concepts are applied. A short introduction to dispersed generation is given and its importance for energy economics today is described. We establish a mixed-integer linear model that represents the operation of such a system, which is run by a German utility. All major technical and economical constraints are included. Computations for single-scenario problems and an expectation-based model show the benefits of stochastic optimization. Then we apply mean-risk optimization and first order dominance constrained programming. Computations with different risk measures and benchmark profiles are reported. The arising deterministic equivalents are solved by standard solvers on the one hand and dual decomposition methods on the other hand to compare the performance of both approaches. The results prove the efficiency and superiority of the decomposition methods, especially for test instances with many different future scenarios.

2 Risk Measures in Two-Stage Stochastic Programs

Whereas the mean describes the expected outcome of a decision, risk measures reflect a decision's riskiness. Since there are various perceptions of risk, many different risk measures are commonly used. One decision maker might appraise a decision very risky if the corresponding outcome shows a high variability. Another might define a personnel ruin level and prefer decisions that likely do not exceed it while random costs falling below his ruin level are accepted. Though we can only present a small collection of risk measures here, we try to cover some different definitions of risk.

In what follows we restrict our considerations to risk measures that fulfill some generally well-accepted properties. Furthermore, these measures are chosen to fit into the described two-stage stochastic optimization scheme. In view of structure and stability, they should provide as good properties as the purely expectation-based model (1.10) and allow for a transfer of the algorithmic methods established for this model below. For instance, they should be structurally sound in the sense that they maintain the mixed-integer linear structure of the problem and algorithmically sound in the sense that they provide a desirable block-structure of the constraint matrix, which we are going to exploit in our algorithm. Details are given in Sections 2.2.1, 2.2.2, and 2.2.3.

From now on assume that $\mathbb{E}(|f(x,z)|) < +\infty$ for every fixed $x \in X$. Then the subsequent collection of risk measures offers the mentioned properties to be "stochastically sound", "structurally sound", and "computationally sound".

2.1 Risk Measures

Risk measures are defined here for the special random variable $f(x,z)$ which represents the first-stage and optimal second-stage costs of a two-stage stochastic optimization problem induced by a decision x and the realization $z(\omega)$ of the probabilistic right hand side. Nevertheless, all risk measures are suitable for arbitrary random variables i.e., $f(x,z)$ could be replaced by a general random variable $X : \Omega \to \mathbb{R}$.

It will be distinguished between risk measures which reflect deviations of the random variable and risk measures which are based on quantiles of the underlying distribution.

2.1.1 Deviation measures

A *deviation measure* requires a preselected threshold $\eta \in \mathbb{R}$, $\eta > 0$ which represents a pain or even a ruin level of the decision maker. Deviation measures describe in a stochastic sense, if or how far the random costs $f(x,z)$ exceed or fall below this threshold.

A very popular deviation measure is the *Variance* of a random variable. The mean of the random variable serves as preselected threshold and then the Variance represents the expected squared deviation of the random variable from its mean (to both sides)

$$Q_V(x) := \mathbb{E}\Big(\big[f(x,z) - \mathbb{E}(f(x,z)) \big]^2 \Big).$$

Markowitz ([Mar52]) and Tobin ([Tob58]) already used the Variance as risk measure for their mean risk analysis applied to portfolio optimization in the 1950's. But the mean-variance model has weak mathematical properties, for example, it is not consistent with the theory of stochastic dominance, which will be introduced in Section 3.1. Markowitz later suggested to replace the Variance by the *Semi-Variance* ([Mar87]), which calculates the expected excess (or shortfall depending on the definition) of the random variable over its mean and shows better properties

$$Q_{SV}(x) := \mathbb{E}\Big(\max\big\{ f(x,z) - \mathbb{E}(f(x,z)), 0 \big\}^2 \Big).$$

Moreover, from an economical point of view the use of the Semi-Variance instead of the Variance is sensible. If we assume that $f(x,z)$ represents the costs resulting from an investment, we want to penalize only those cases where $f(x,z)$ exceeds its mean, but not the desirable cases where it even drops below the expected costs.

Despite its favorable characteristics from a mathematical and practical point of view, the Semi-Variance does not show the computational properties we want to exploit in our algorithmic approach described in Section 2.2.3. Therefore, we prefer the so-called *Expected Excess* as deviation measure which reflects the expectation of the portion of costs $f(x,z)$ exceeding a given target $\eta \in \mathbb{R}$

$$Q_{EE_\eta}(x) := \mathbb{E}\Big(\max\big\{ f(x,z) - \eta, 0 \big\} \Big).$$

2.1.2 Quantile-based risk measures

Another class of risk measures uses the quantiles of the cumulative distribution function of a random variable. Let $0 < q < 1$ and X a continuously distributed, real-valued random variable then the *q-quantile* of X is defined as the unique value $\xi \in \mathbb{R}$ with

$$\mathbb{P}(X \leq \xi) = q.$$

A more general definition which is suitable for arbitrary random variables e.g., not continuously, but discretely distributed ones, is the following: Let $0 < q < 1$ and X an arbitrary, real-valued random variable then the q-quantile of X is the value $\xi \in \mathbb{R}$ with

$$\mathbb{P}(X \leq \xi) \geq q \quad \text{and} \quad \mathbb{P}(X \geq \xi) \geq 1 - q.$$

Simply spoken, a quantile divides the space of outcomes of the random variable into parts with respect to its probability distribution. The most "famous" quantile of a random variable is the median (0.5-quantile of X) that divides the space of outcomes into two parts both having a probability of 0.5.

We want to use three prominent quantile-based risk measures:

- The *Excess Probability* calculates the probability that the costs induced by the first-stage decision x and the second-stage decision $y = y(x, z)$ exceed a given target $\eta \in \mathbb{R}$

$$\mathbb{Q}_{EP_\eta}(x) := \mathbb{P}\Big(\{\omega \in \Omega : f(x, z) > \eta\}\Big).$$

  Let $F_x(\eta) := \mathbb{P}\Big(\{\omega \in \Omega : f(x, z) \leq \eta\}\Big)$ denote the cumulative distribution function of $f(x, z)$ for a fixed $x \in X$ then the Excess Probability can also be defined by

$$\mathbb{Q}_{EP_\eta}(x) := 1 - F_x(\eta).$$

- The *α-Value-at-Risk*, which is well-known from financial applications, gives the minimum costs occurring in the $(1 - \alpha) \cdot 100\%$ worst cases. It can be defined using the above introduced Excess Probability

$$\mathbb{Q}_{VaR_\alpha}(x) := \min\Big\{\eta : \mathbb{Q}_{EP_\eta}(x) \leq 1 - \alpha, \eta \in \mathbb{R}\Big\}.$$

  Since this definition does not obviously show that the α-Value-at-Risk is a quantile-based risk measure, another more explicit definition shall be given

$$\mathbb{Q}_{VaR_\alpha}(x) := \min\Big\{\eta : F_x(\eta) \geq \alpha, \eta \in \mathbb{R}\Big\},$$

where F_x again is the cumulative distribution function of $f(x,z)$ for a fixed $x \in X$.

- Whereas the α-Value-at-Risk reflects the costs of the cheapest of the $(1 - \alpha) \cdot 100\%$ worst scenarios, the α-*Conditional Value-at-Risk* calculates the expectation of costs of all $(1 - \alpha) \cdot 100\%$ worst cases. For a continuously distributed $f(x,z)$ there is a very intuitive formula for the α-Conditional Value-at-Risk using the above defined α-Value-at-Risk:

  If there is no probability atom at $\mathbb{Q}_{VaR_\alpha}$, and so $F_x(\mathbb{Q}_{VaR_\alpha}(x)) = \alpha$, the α-Conditional Value-at-Risk is equal to the conditional expectation

  $$\mathbb{Q}_{CVaR_\alpha}(x) := \mathbb{E}\big(f(x,z) \,|\, f(x,z) \geq \mathbb{Q}_{VaR_\alpha}(x)\big). \qquad (2.1)$$

In contrast, if there is no η such that $F_x(\eta) = \alpha$, which may occur for discretely distributed random variables, (2.1) is not the correct definition for the α-Conditional Value-at-Risk. Then the following general definition holds

$$\mathbb{Q}_{CVaR_\alpha}(x) := \text{mean of the } \alpha-\text{tail distribution of } f(x,z),$$

where the cumulative distribution function belonging to this α-tail distribution is defined by

$$F_x^\alpha(\eta) := \left\{ \begin{array}{cl} 0 & \text{for } \eta < \mathbb{Q}_{VaR_\alpha}(x), \\ (F_x(\eta) - \alpha)/(1 - \alpha) & \text{for } \eta \geq \mathbb{Q}_{VaR_\alpha}(x). \end{array} \right.$$

Finally, we end up with another general definition of the α-Conditional Value-at-Risk which can be deduced from the above one (see [RU00, ST06]) and which we are going to use for further analysis

$$\mathbb{Q}_{CVaR_\alpha}(x) := \min\left\{ \eta + \frac{1}{1 - \alpha} \mathbb{Q}_{EE_\eta}(x) \,:\, \eta \in \mathbb{R} \right\}.$$

2.2 Mean-Risk Models

The described risk measures can be used to improve the expectation-based optimization problem including an additional risk term. Using the weighted sum of the expected costs $\mathbb{Q}_{\mathbb{E}}$ and a risk measure $\mathbb{Q}_{\mathscr{R}}$ as objective, we obtain the so-called *mean-risk model*

$$\min\left\{ \mathbb{Q}_{MR}(x) \,:\, x \in X \right\} \qquad (2.2)$$

with

$$Q_{MR}(x) := Q_{\mathbb{E}}(x) + \rho \cdot Q_{\mathscr{R}}(x), \quad \rho > 0 \text{ fixed.}$$

The chosen decision $x \in X$ then has to be optimal for both – usually competitive – characteristics. A notion of optimality can be derived from multi-objective optimization if we consider the mean-risk model (2.2) as a scalarization of the multi-objective optimization problem

$$\min \left\{ \left(Q_{\mathbb{E}}(x), Q_{\mathscr{R}}(x) \right) : x \in X \right\}. \tag{2.3}$$

A decision $\bar{x} \in X$ is called *efficient* for (2.3) if there exists no other $x \in X$ such that $Q_{\mathbb{E}}(x) \le Q_{\mathbb{E}}(\bar{x})$ as well as $Q_{\mathscr{R}}(x) \le Q_{\mathscr{R}}(\bar{x})$ and at least one inequality is strict. All efficient points form the *efficient frontier* of a multi-objective optimization problem. It holds that every optimal solution of (2.2) with a fixed ρ is an efficient point for (2.3). However, we cannot detect the whole efficient frontier solving the mean-risk model because of the integer requirements in the second stage and the resulting non-convexity of the problem. With various weights ρ we can only trace the so-called supported part of the efficient frontier. For more details about multi-objective optimization see [Ben01, Ehr00, Kor01].

Below, we first describe structural and stability properties of (2.2). Then deterministic equivalents are presented, which arise if a discretization of the probability distribution of $z(\omega)$ is considered. Finally, we give a tailored algorithm based on Lagrangian relaxation that is embedded into a branch-and-bound scheme.

2.2.1 Results concerning structure and stability

This section deals with results concerning structure and stability of mean-risk optimization problems (2.2) with respect to perturbations of the underlying probability measure. Such results can be deduced considering the mean-risk models as special nonlinear parametric optimization problems. We start with some basic characteristics of the second-stage value function Φ. A second paragraph concerns a specified standard proposition from parametric optimization. Finally, we show exemplarily how to apply this proposition to the expectation-based problem as well as the pure risk model with α-Conditional Value-at-Risk and obtain the desired stability results. Also the well-posedness of the optimization problems will be considered.

We will confine ourselves to a review of the most important results without giving proofs. For further details references are mentioned.

The motivation for stability analysis is given by two aspects occurring in practice. The first aspect is the fact that usually the underlying probability distribution of the uncertain input data is not completely known. Only a discrete approximation is used, which often results from subjective considerations as well as expert knowledge and experience. The second aspect concerns the structure of $\mathbb{Q}_{\mathbb{E}}(x)$ and therewith the structure of the mean-risk models. The calculation of $\mathbb{Q}_{\mathbb{E}}(x)$ requires the solution of a multivariate integration in high dimension relying on a continuous integrating measure which also necessitates a discrete approximation. The question arises, what effects on the optimal solution of a mean-risk model we must expect if we use an approximation of the probability measure instead of the exact measure. In other words, the aim of the stability analysis is to make sure that small perturbations of the underlying probability measure entail only small perturbations of an optimal solution.

Hence, the presented stability results constitute the basis for the computations done with mean-risk models, which will be presented later on. Furthermore, there are parallels to the proofs of stability results for dominance constrained problems being discussed in detail in subsequent sections.

The first publications on stability of stochastic optimization problems date back to the 1970's. In [Ber75, Kan78] the notion of stability in stochastic programming is already used and in [Kal74, Kan77, Mar75, Mar79, Ols76, Wet79] first results on approximation in stochastic programming are presented. Two recent surveys on stability of stochastic programs with respect to perturbations of the underlying probability measure are [Röm03, Sch00].

Below, assumptions not on all second-stage variables y, but on the integer ones among them are formulated. Hence, in this section let $y = (y', y'')$, $W = (W', W'')$, and $q = (q', q'')$ and use the reformulation

$$\min \left\{ c^\top x + \Phi(z(\omega) - Tx) : x \in X \right\}$$

with

$$\Phi(t) := \min \left\{ q'^\top y' + q''^\top y'' : W'y' + W''y'' = t, \, y' \in \mathbb{Z}_+^{\bar{m}}, \, y'' \in \mathbb{R}_+^{m'} \right\} \quad (2.4)$$

of a two-stage stochastic optimization problem.

For the following qualitative stability results we have to endow the space $\mathscr{P}(\mathbb{R}^s)$ with some notion of convergence. An appropriate choice is *weak convergence of probability measures* which covers many other important modes of convergence (for reference see [Bil68, Dud89, VdVW96]). A sequence (μ_n) in $\mathscr{P}(\mathbb{R}^s)$ converges weakly to $\mu \in \mathscr{P}(\mathbb{R}^s)$, written $\mu_n \xrightarrow{w} \mu$, if for any bounded continuous function $h : \mathbb{R}^s \to \mathbb{R}$ holds that

$$\int_{\mathbb{R}^s} h(z) \mu_n(dz) \to \int_{\mathbb{R}^s} h(z) \mu(dz) \quad \text{as} \quad n \to \infty.$$

In finite dimensions weak convergence of probability measures coincides with the pointwise convergence of distribution functions at all continuity points of the limiting distribution function.

For quantitative results we need an appropriate notion of distance of probability measures. Such a distance measure is given by the *variational distance* $\sigma_{\mathscr{B}_\mathscr{K}}$ of probability measures on $\mathscr{P}(\mathbb{R}^s)$. Assume that $\mathscr{B}_\mathscr{K}$ denotes the class of all (closed) bounded polyhedra in $\mathbb{R}^s$, each of whose facets parallels a facet of $\mathscr{K} := \operatorname{pos}W''$ or a facet of $\{z \in \mathbb{R}^s : ||z||_\infty \leq 1\}$, where $||z||_\infty = \max_{i=1,\ldots,s} |z_i|$. Then we define $\sigma_{\mathscr{B}_\mathscr{K}}$ by

$$\sigma_{\mathscr{B}_\mathscr{K}}(\mu, \nu) := \sup \left\{ |\mu(B) - \nu(B)| : B \in \mathscr{B}_\mathscr{K} \right\}.$$

As described in [Sch96], this variational distance is a metric.

Properties of the value function Φ Basically, some structural characteristics of the value function Φ are needed, which are gained from parametric optimization. We already mentioned the lower semicontinuity of Φ under the assumptions (1.5) – (1.7) in Section 1.1.1. Besides, some more characteristics can be stated.

Proposition 2.1
Assume (1.5) – (1.7) then it holds that

(i) Φ is real-valued and lower semicontinuous on $\mathbb{R}^s$;

(ii) there exists a countable partition $\mathbb{R}^s = \bigcup_{i=1}^\infty \mathscr{T}_i$ such that the restrictions of Φ to $\mathscr{T}_i$ are piecewise linear and Lipschitz continuous with a uniform constant not depending on i, and on each $\mathscr{T}_i$ the function Φ admits a representation

$$\Phi(t) = \min_{y' \in Y(t)} \left\{ q^\top y' + \max_{k=1,\ldots,K} d_k^\top (t - W'y') \right\},$$

where $Y(t) := \left\{ y' \in \mathbb{Z}_+^{\bar{m}} : t \in W'y' + W''(\mathbb{R}_+^{m'}) \right\}$ and d_k, $k = 1,\ldots,K$, are the vertices of the polyhedron $\{u \in \mathbb{R}^s : W''^\top u \leq q''\}$;

(iii) each of the sets $\mathscr{T}_i$ has a representation $\mathscr{T}_i = \{t_i + \mathscr{K}\} \setminus \bigcup_{j=1}^N \{t_{ij} + \mathscr{K}\}$, where $\mathscr{K}$ denotes the polyhedral cone $W''(\mathbb{R}_+^{m'})$ and t_i, t_{ij} are suitable points from $\mathbb{R}^s$; moreover N does not depend on i;

(iv) there exist positive constants β, γ such that $|\Phi(t_1) - \Phi(t_2)| \leq \beta ||t_1 - t_2|| + \gamma$ whenever $t_1, t_2 \in \mathbb{R}^s$.

These characteristics will also play a role in the stability analysis of the first order dominance constrained stochastic optimization problems, which is addressed later in this thesis.

In a second step we collect some basics from nonlinear parametric optimization.

Results from parametric optimization   We consider stochastic optimization problems as a special case of a nonlinear parametric optimization problem

$$(P_\nu) \qquad \min \big\{ \mathscr{F}(x,\nu) \,:\, x \in \mathscr{X}(\nu) \big\}.$$

The objective function $\mathscr{F}$ as well as the constraint set $\mathscr{X}(\nu) := \big\{ x \in \mathbb{R}^m \,:\, \mathscr{H}(x,\nu) \leq 0 \big\}$ depend on the probability measure ν, which serves as the variational parameter. Furthermore, let

$$\varphi(\nu) \quad := \quad \inf \big\{ \mathscr{F}(x,\nu) \,:\, x \in \mathscr{X}(\nu) \big\}$$

and

$$\psi(\nu) \quad := \quad \operatorname{argmin} \big\{ \mathscr{F}(x,\nu) \,:\, x \in \mathscr{X}(\nu) \big\}$$

denote the *optimal value function* and the *solution set mapping* of the parametric problem (P_ν), respectively. To derive propositions about the stability of a mean-risk optimization problem with perturbed underlying probability measure, we now have to answer the question how a perturbation of ν affects $\varphi(\nu)$ and $\psi(\nu)$.

There are several methods known in parametric optimization to prove properties concerning the continuity of the optimal value function and the solution set mapping of a problem. We follow the approach described in [BGK$^+$82, Ber63, Dan67, DFS67, Kla87], which presumes some properties of the objective function $\mathscr{F}$ and the constraints $\mathscr{H}(x,\nu) \leq 0$ to deduce the desired properties of $\varphi(\nu)$ and $\psi(\nu)$.

The following lemma is a specialized version of some standard result from parametric optimization that can be found in [BGK$^+$82]. Applying this lemma to mean-risk models – considered as parametric optimization problems – will lead to the stability results we aim at.

Lemma 2.2
Consider the problem

$$(P_\lambda) \qquad \inf \big\{ \mathscr{F}(\theta,\lambda) \,:\, \theta \in \Theta \big\}, \quad \lambda \in \Lambda,$$

where $\mathscr{F} : \Theta \times \Lambda \to \mathbb{R}$ and Λ, Θ are metric spaces endowed with some notion of convergence. Let $\varphi(\lambda)$ and $\psi(\lambda)$ denote the infimum and the optimal set of (P_λ), respectively. Then the following holds:

(i) φ is lower semicontinuous at $\lambda_0 \in \Lambda$ if $\mathscr{F}$ is lower semicontinuous on $\Theta \times \{\lambda_0\}$ and if for each sequence $\lambda_n \to \lambda_0$ there exists a compact subset K of Θ such that $\psi(\lambda_n) \cap K \neq \varnothing$ holds for all $n \in \mathbb{N}$;

(ii) φ is upper semicontinuous at λ_0 if $\mathscr{F}(\theta, .)$ is upper semicontinuous at λ_0 for all $\theta \in \Theta$;

(iii) let $\lambda_1, \lambda_2 \in \Lambda$ and assume that there exist $\theta_i \in \psi(\lambda_i)$, $i = 1, 2$. Then it holds that

$$|\varphi(\lambda_1) - \varphi(\lambda_2)| \leq \max_{i=1,2} \left\{ |\mathscr{F}(\theta_i, \lambda_1) - \mathscr{F}(\theta_i, \lambda_2)| \right\}.$$

The application of Lemma 2.2 to the pure expectation-based problem is described in the following paragraph.

Well-posedness and stability of the pure expectation-based problem Recall the pure expectation-based optimization problem

$$\min \left\{ \mathbb{Q}_{\mathbb{E}}(x, \mu) \; : \; x \in X \right\} \tag{2.5}$$

with

$$\mathbb{Q}_{\mathbb{E}}(x, \mu) := \int_{\mathbb{R}^s} c^\top x + \Phi(z - Tx) \, \mu(dz)$$

and

$$\mu := \mathbb{P} \circ z^{-1},$$

the probability measure induced by $z(\omega)$.

First of all, we collect some structural characteristics of problem (2.5). For convenience let us define

$$M_d(x) := \left\{ z \in \mathbb{R}^s \; : \; \Phi \text{ not continuous at } z - Tx \right\}. \tag{2.6}$$

We can conclude that the pure expectation-based problem is well-posed in the sense that the infimum is finite and attained. This follows from the lower semicontinuity of the problem's objective function in combination with a nonempty and compact constraint set.

Proposition 2.3
Assume (1.5) – (1.7), and let $\mu \in \mathscr{P}(\mathbb{R}^s)$ such that $\int_{\mathbb{R}^s} ||z|| \mu(dz) < \infty$. Then $\mathbb{Q}_{\mathbb{E}} : \mathbb{R}^m \to \mathbb{R}$, $\mathbb{Q}_{\mathbb{E}}(x) := \int_{\mathbb{R}^s} c^\top x + \Phi(z - Tx) \mu(dz)$ is real-valued and lower semicontinuous. If in addition $\mu(M_d(x)) = 0$, then $\mathbb{Q}_{\mathbb{E}}(x)$ is continuous at x.

For a proof see [Sch95].

For the next two propositions, which provide the assumptions needed to apply
Lemma 2.2 to problem (2.5), we have to narrow the space of Borel probability
measures $\mathscr{P}(\mathbb{R}^s)$ to a subspace $\Delta_{p,C}(\mathbb{R}^s)$, which is defined as

$$\Delta_{p,C}(\mathbb{R}^s) := \left\{ v \in \mathscr{P}(\mathbb{R}^s) : \int_{\mathbb{R}^s} ||z||^p \, v(dz) \leq C \right\}$$

with $p, C \in \mathbb{R}$, $p > 1$, $C > 0$.

The first proposition, a qualitative one, concerns the continuity of $\mathbb{Q}_{\mathbb{E}}(x, \mu)$ together in x and μ.

Proposition 2.4
Assume (1.5) – (1.7). Moreover, let $x \in \mathbb{R}^m$ and $\mu \in \Delta_{p,C}(\mathbb{R}^s)$ with arbitrary $p > 1$ and
$C > 0$ such that $\mu(M_d(x)) = 0$. Then $\mathbb{Q}_{\mathbb{E}} : \mathbb{R}^m \times \Delta_{p,C}(\mathbb{R}^s) \to \mathbb{R}$ is continuous at (x, μ).

The second proposition, a quantitative one, gives an upper bound on the varia-
tion of $\mathbb{Q}_{\mathbb{E}}(x, \mu)$ if instead of μ a probability measure "near by" μ, in the sense of
the variational distance, is used.

Proposition 2.5
Assume (1.5) – (1.7), and let $p > 1$, $C > 0$, and $D \subseteq \mathbb{R}^m$ be nonempty and bounded. Then
there exist constants $L > 0$ and $\delta > 0$ such that

$$\sup_{x \in D} |\mathbb{Q}_{\mathbb{E}}(x, \mu) - \mathbb{Q}_{\mathbb{E}}(x, v)| \leq L \cdot \sigma_{\mathscr{B}_{\mathscr{K}}}(\mu, v)^{\frac{p-1}{p(s+1)}}$$

whenever $\mu, v \in \Delta_{p,C}(\mathbb{R}^s)$, $\sigma_{\mathscr{B}_{\mathscr{K}}}(\mu, v) < \delta$.

Proofs are given in [Sch95, Sch96].

Now consider the pure expectation-based problem (2.5) as a parametric optimization problem with parameter μ

$$(P_\mu) \qquad \min \left\{ \mathbb{Q}_{\mathbb{E}}(x, \mu) : x \in X \right\}$$

with the localized optimal value function and solution set mapping

$$\varphi_V(\mu) := \inf \left\{ \mathbb{Q}_E(x, \mu) : x \in X \cap \mathrm{cl} V \right\},$$
$$\psi_V(\mu) := \left\{ x \in X \cap \mathrm{cl} V : \mathbb{Q}_E(x, \mu) = \varphi_V(\mu) \right\},$$

where $V \subseteq \mathbb{R}^m$.

We use these localized versions of the definitions because $\mathbb{Q}_{\mathbb{E}}(x, \mu)$ usually is not convex in x, which means that local minimizers are not always global minimizers. To avoid pathologies, we apply the notion of a *complete local minimizing (CLM) set*: Given $\mu \in \mathscr{P}(\mathbb{R}^s)$, a nonempty set $Z \subseteq \mathbb{R}^m$ is called a CLM set of (P_μ) with respect to some open set $V \subseteq \mathbb{R}^m$ if $Z = \psi_V(\mu) \subseteq V$ (for more details see [Rob87]).

Using Lemma 2.2, we arrive at the following two stability results (again a qualitative and a quantitative one) for the pure expectation-based optimization problem.

Proposition 2.6
Assume (1.5) – (1.7). Let $\mu \in \Delta_{p,C}(\mathbb{R}^s)$ for arbitrary $p > 1$, $C > 0$ such that $\mu(M_d(x)) = 0$ for all $x \in X$. Furthermore, suppose that there exists a subset $Z \subseteq \mathbb{R}^m$ which is a CLM set for (P_μ) with respect to some bounded open set $V \subseteq \mathbb{R}^m$. Then it holds that

(i) the function $\varphi_V : \Delta_{p,C}(\mathbb{R}^s) \to \mathbb{R}$ is continuous at μ;

(ii) the multifunction $\psi_V : \Delta_{p,C}(\mathbb{R}^s) \to 2^{\mathbb{R}^m}$ is Berge upper semicontinuous at μ i.e., for any open set $\mathscr{O} \subseteq \mathbb{R}^m$ with $\psi_V(\mu) \subseteq \mathscr{O}$ there exists a neighborhood $\mathscr{N}$ of μ in $\Delta_{p,C}(\mathbb{R}^s)$ such that $\psi_V(\nu) \subseteq \mathscr{O}$ for all ν in $\mathscr{N}$;

(iii) there exists a neighborhood $\mathscr{N}'$ of μ in $\Delta_{p,C}(\mathbb{R}^s)$ such that for all ν in $\mathscr{N}'$ the set $\psi_V(\nu)$ is a CLM set for (P_ν) with respect to V. In particular, this implies that $\psi_V(\nu)$ is a nonempty set of local minimizers whenever $\nu \in \mathscr{N}'$.

Proposition 2.7
Assume (1.5) – (1.7), let $\mu \in \Delta_{p,C}(\mathbb{R}^s)$ for arbitrary $p > 1$ and $C > 0$, and let $V \subseteq \mathbb{R}^m$ be bounded. Then there exist constants $L > 0$ and $\delta > 0$ such that

$$|\varphi_V(\mu) - \varphi_V(\nu)| \le L \cdot \sigma_{\mathscr{B}_{\mathscr{K}}}(\mu, \nu)^{\frac{p-1}{p(s+1)}}$$

whenever $\nu \in \Delta_{p,C}(\mathbb{R}^s)$, $\sigma_{\mathscr{B}_{\mathscr{K}}}(\mu, \nu) < \delta$.

For proofs see [Sch95, Sch96, Tie05].

As mentioned before, the proofs mainly make use of Lemma 2.2 and Propositions 2.3, 2.4, and 2.5, as well as the assumption that the solution set of the unperturbed problem is nonempty and compact. This is represented by the boundedness of V in Propositions 2.6 and 2.7. For further details on structural properties of the pure expectation-based problem we refer to some standard text books on stochastic optimization like [BL97, KW94, Pré95, RS03].

Following a similar scheme, stability results can be deduced for the pure risk model with α-Conditional Value-at-Risk. They are reviewed in the following paragraph.

Well-posedness and stability of the pure risk model with α-Conditional Value-at-Risk   Recall the pure risk model using the α-Conditional Value-at-Risk as risk measure

$$\min \left\{ \mathbb{Q}_{CVaR}(\alpha,x,\mu) \ : \ x \in X \right\}$$

with

$$\mathbb{Q}_{CVar}(\alpha,x,\mu) := \min \left\{ \eta + \frac{1}{1-\alpha} \mathbb{Q}_{EE}(\eta,x,\mu) \ : \ \eta \in \mathbb{R} \right\},$$

$\mu \in \mathscr{P}(\mathbb{R}^s)$ the probability measure induced by $z(\omega)$, $\alpha \in (0,1)$ a preselected, fixed parameter, and $\eta \in \mathbb{R}$ the threshold for the deviation measure $\mathbb{Q}_{EE_\eta}$.

The first proposition states the well-posedness of this optimization problem assumed that the constraint set is nonempty and compact.

Proposition 2.8
Assume (1.5) – (1.7), $\mu \in \mathscr{P}(\mathbb{R}^s)$ with $\int_{\mathbb{R}^s} ||z|| \mu(dz) < \infty$, and let q', q'' be rational vectors. Then $\mathbb{Q}_{CVaR} : (0,1) \times \mathbb{R}^m \to \mathbb{R}$ is real-valued and lower semicontinuous. If in addition $\mu(M_d(x)) = 0$, then $\mathbb{Q}_{CVaR}(\alpha,x)$ is continuous at (α,x) for all $\alpha \in (0,1)$.

A proof can be found in [Tie05].

The following two propositions show that the assumptions of Lemma 2.2 are fulfilled for $\mathbb{Q}_{CVaR}$.

Proposition 2.9
Assume (1.5) – (1.7) and that q', q'' are rational vectors. Let $x \in \mathbb{R}^m$ as well as $\mu \in \Delta_{p,C}(\mathbb{R}^s)$ with arbitrary $p > 1$ and $C > 0$ such that $\mu(M_d(x)) = 0$. Then $\mathbb{Q}_{CVaR} : (0,1) \times \mathbb{R}^m \times \Delta_{p,C}(\mathbb{R}^s) \to \mathbb{R}$ is continuous at (α,x,μ) for all $\alpha \in (0,1)$.

Proposition 2.10
Assume (1.5) – (1.7) and that q', q'' are rational vectors. Let $\mu \in \Delta_{p,C}(\mathbb{R}^s)$ with arbitrary $p > 1$ and $C > 0$ and $D \subseteq \mathbb{R}^m$ be nonempty and bounded. Then there exist constants $L > 0$ and $\delta > 0$ such that

$$\sup_{x \in D} |\mathbb{Q}_{CVaR}(x,\mu) - \mathbb{Q}_{CVaR}(x,\nu)| \leq L \cdot \sigma_{\mathscr{B}_{\mathscr{K}}}(\mu,\nu)^{\frac{p-1}{p(s+2)}}$$

whenever $\nu \in \Delta_{p,C}(\mathbb{R}^s)$, $\sigma_{\mathscr{B}_{\mathscr{K}}}(\mu,\nu) < \delta$.

For proofs see [Tie05].

We now consider the pure risk problem with α-Conditional Value-at-Risk as a parametric optimization problem with parameter μ

$$(P_\mu) \qquad \min \left\{ \mathbb{Q}_{CVaR}(\alpha,x,\mu) \ : \ x \in X \right\}$$

with the localized optimal value function and solution set mapping on a subset $V \subseteq \mathbb{R}^m$

$$\varphi_V(\mu) \; := \; \inf \big\{ \mathbb{Q}_{CVaR}(\alpha, x, \mu) \; : \; x \in X \cap \mathrm{cl}\, V \big\},$$
$$\psi_V(\mu) \; := \; \big\{ x \in X \cap \mathrm{cl}\, V \; : \; \mathbb{Q}_{CVaR}(\alpha, x, \mu) = \varphi_V(\mu) \big\}.$$

Finally, Lemma 2.2 together with Propositions 2.9 and 2.10 yields the following qualitative and quantitative stability results.

Proposition 2.11
Assume (1.5) – (1.7) and that q', q'' are rational vectors. Furthermore, let $\mu \in \Delta_{p,C}(\mathbb{R}^s)$ for arbitrary $p > 1$ and $C > 0$ such that $\mu(M_d(x)) = 0$ for all $x \in X$. Suppose that there exists a subset $Z \subseteq \mathbb{R}^m$ which is a CLM set for (P_μ) with respect to some bounded open set $V \subseteq \mathbb{R}^m$. Then it holds that

(i) the function $\varphi_V : \Delta_{p,C}(\mathbb{R}^s) \to \mathbb{R}$ is continuous at μ;

(ii) the multifunction $\psi_V : \Delta_{p,C}(\mathbb{R}^s) \to 2^{\mathbb{R}^m}$ is Berge upper semicontinuous at μ i.e., for any open set $\mathcal{O} \subseteq \mathbb{R}^m$ with $\psi_V(\mu) \subseteq \mathcal{O}$ there exists a neighborhood $\mathcal{N}$ of μ in $\mathscr{P}(\mathbb{R}^s)$ such that $\psi_V(v) \subseteq \mathcal{O}$ for all v in $\mathcal{N}$;

(iii) there exists a neighborhood $\mathcal{N}'$ of μ in $\Delta_{p,C}(\mathbb{R}^s)$ such that for all v in $\mathcal{N}'$ the set $\psi_V(v)$ is a CLM set for (P_v) with respect to V. In particular, this implies that $\psi_V(v)$ is a nonempty set of local minimizers whenever $v \in \mathcal{N}'$.

Proposition 2.12
Assume (1.5) – (1.7) and that q', q'' are rational vectors. Furthermore, let $\mu \in \Delta_{p,C}(\mathbb{R}^s)$ for arbitrary $p > 1$ and $C > 0$, and let $V \subseteq \mathbb{R}^m$ be bounded. Then there exist constants $L > 0$ and $\delta > 0$ such that

$$| \varphi_V(\mu) - \varphi_V(v) | \leq L \cdot \sigma_{\mathscr{B}_{\mathscr{K}}}(\mu, v)^{\frac{p-1}{p(s+2)}}$$

for all $v \in \Delta_{p,C}(\mathbb{R}^s)$, $\sigma_{\mathscr{B}_{\mathscr{K}}}(\mu, v) < \delta$.

Detailed deductions and proofs can be found in [ST06, Tie05].

In a similar way, the well-posedness and stability under perturbations of the underlying probability measure can be deduced for the risk models with Expected Excess and Excess Probability. Only for the α-Value-at-Risk a quantitative stability result cannot be proven following the above scheme. This is addressed in [Tie05] and shall not be discussed in detail here.

2.2.2 Deterministic equivalents

In this section deterministic equivalents of the above introduced pure expectation-based model and pure risk models are presented. They arise if we replace the exact probability measure induced by the stochastic right hand side $z(\omega)$ by an appropriate approximation. Deterministic reformulations are needed to make the continuous mean-risk optimization problems tractable by mathematical algorithms.

The stability results of Section 2.2.1 serve as justification for the usage of deterministic equivalents presumed that the approximative, discrete probability measure converges weakly to the exact underlying probability measure. Since several important specific modes of convergence are covered by weak convergence, different *approximation schemes* can be applied to find a suitable approximation. For example, discretizations of probability measures via almost surely converging densities (Scheffé's Theorem [Bil68]), via conditional expectations (see [BW86, Kal87]), or via estimation using empirical measures (Glivenko-Cantelli almost sure uniform convergence [Pol84]) often produce weakly converging sequences of discrete probability measures.

Using such suitable approximations guarantees that the optimal solution of a deterministic equivalent converges to the optimal solution of the exact stochastic mean-risk model.

Recall the two-stage stochastic optimization problem (1.4)

$$\min_x \left\{ c^\top x + \min_y \left\{ q^\top y \,:\, Wy = z(\omega) - Tx, \, y \in \mathbb{Z}_+^{\bar{m}} \times \mathbb{R}_+^{m'} \right\} \,:\, x \in X \right\}.$$

Moreover, assume that the random variable $z(\omega)$ is discretely distributed with finitely many realizations $z_1, \ldots, z_L$, also called scenarios, and corresponding probabilities $\pi_1, \ldots, \pi_L$.

The deterministic equivalent for the pure expectation-based model $\min\{\mathbb{Q}_\mathbb{E}(x) : x \in X\}$ then reads

$$\min \left\{ c^\top x + \sum_{l=1}^{L} \pi_l q^\top y_l \,:\, \begin{array}{rcll} Tx + Wy_l &=& z_l & \forall l \\ x \in X, \ y_l \in \mathbb{Z}_+^{\bar{m}} \times \mathbb{R}_+^{m'} & & & \forall l \end{array} \right\}. \tag{2.7}$$

Whereas the first-stage decision x remains nonanticipative (it does not foresee the future), the second-stage decision y is represented by variables $y_1, \ldots, y_L$, hence can be different for each occurring scenario.

We continue with deterministic equivalents for the pure risk models including Excess Probability, Expected Excess, α-Value-at-Risk and α-Conditional Value-at-Risk. Combining them with model (2.7), yields deterministic equivalents for the mean-risk model (2.2).

Pure risk model with Excess Probability Recall that the *support* supp μ of $\mu \in \mathscr{P}(\mathbb{R}^s)$ is the smallest closed subset of $\mathbb{R}^s$ with μ-measure 1. To formulate the deterministic equivalent of the risk model using the Excess Probability as risk measure, we need the following result.

Proposition 2.13
Assume (1.5) – (1.7), that μ has bounded support, that X is bounded, and let $\eta \in \mathbb{R}$. Then there exists a constant M such that

$$M > \sup \left\{ c^\top x + \Phi(z - Tx) \; : \; z \in \operatorname{supp}\mu, \, x \in X \right\}.$$

Furthermore, it holds that the problems

$$\min \left\{ \mathbb{Q}_{EP_\eta}(x) \; : \; x \in X \right\} \quad \text{and} \quad \min \left\{ \tilde{E}_\eta(x) \; : \; x \in X \right\}$$

with

$$\tilde{E}_\eta(x) \; := \; \int_{\mathbb{R}^s} \tilde{\Phi}(z - Tx, c^\top x - \eta) \, \mu(dz)$$

and

$$\tilde{\Phi}(t_1, t_2) := \min \left\{ u \; : \; \begin{array}{rcl} Wy & = & t_1 \\ -q^\top y + (M - \eta) \cdot u & \geq & t_2 \\ y \in \mathbb{Z}_+^{\bar{m}} \times \mathbb{R}_+^{m'}, \; u \in \{0,1\} & & \end{array} \right\}$$

are equivalent.

For a proof see [Tie05].

Therewith we obtain:

Corollary 2.14
Assume (1.5) – (1.7), that $z(\omega)$ is discretely distributed with finitely many realizations $z_1, \ldots, z_L$ and probabilities $\pi_1, \ldots, \pi_L$ (which yields a bounded support of μ), that $X \subseteq \mathbb{R}^m$ is nonempty as well as compact, and let $\eta \in \mathbb{R}$. Then there exists a constant $M > 0$ such that the stochastic program

$$\min \left\{ \mathbb{Q}_{EP_\eta}(x) \; : \; x \in X \right\}$$

can be equivalently restated as

$$\min \left\{ \sum_{l=1}^{L} \pi_l u_l \ : \ \begin{array}{lll} Wy_l + Tx & = z_l & \forall l \\ c^\top x + q^\top y_l - Mu_l & \leq \eta & \forall l \\ x \in X, \ y_l \in \mathbb{Z}_+^{\bar{m}} \times \mathbb{R}_+^{m'}, \ u_l \in \{0,1\} & & \forall l \end{array} \right\}. \qquad (2.8)$$

Pure risk model with Expected Excess   For the pure risk model using the Expected Excess as risk measure the following proposition holds.

Proposition 2.15
Assume (1.5) – (1.7), that $z(\omega)$ is discretely distributed with finitely many realizations $z_1,\ldots,z_L$ and probabilities $\pi_1,\ldots,\pi_L$, and let $\eta \in \mathbb{R}$. Then the stochastic program

$$\min \left\{ \mathbb{Q}_{EE_\eta}(x) \ : \ x \in X \right\}$$

can be equivalently restated as

$$\min \left\{ \sum_{l=1}^{L} \pi_l v_l \ : \ \begin{array}{lll} Wy_l + Tx & = z_l & \forall l \\ c^\top x + q^\top y_l - v_l & \leq \eta & \forall l \\ x \in X, \ y_l \in \mathbb{Z}_+^{\bar{m}} \times \mathbb{R}_+^{m'}, \ v_l \in \mathbb{R}_+ & & \forall l \end{array} \right\}. \qquad (2.9)$$

Pure risk model with α-Value-at-Risk   For the deterministic equivalent of the pure risk model using α-Value-at-Risk as risk measure the above Proposition 2.13 is needed, too. It states that there exists a parameter $M > 0$ which covers the single-scenario costs of every scenario that can occur. Together with this assumption the following proposition holds.

Proposition 2.16
Assume (1.5) – (1.7), that $z(\omega)$ is discretely distributed with finitely many realizations $z_1,\ldots,z_L$ and probabilities $\pi_1,\ldots,\pi_L$ (which yields a bounded support of μ), that $X \subseteq \mathbb{R}^m$ is nonempty and compact, and let $\alpha \in (0,1)$. Then there exists a constant $M > 0$ such that the stochastic program

$$\min \left\{ \mathbb{Q}_{VaR_\alpha}(x) \ : \ x \in X \right\}$$

can be equivalently restated as

$$
\min \left\{ \eta \; : \quad
\begin{aligned}
Wy_l + Tx &= z_l && \forall l \\
c^\top x + q^\top y_l - Mu_l &\leq \eta && \forall l \\
\sum_{l=1}^{L} \pi_l u_l &\leq 1 - \alpha \\
x \in X, \; y_l \in \mathbb{Z}_+^{\bar{m}} \times \mathbb{R}_+^{m'}, \; u_l &\in \{0,1\}, \; \eta \in \mathbb{R} && \forall l
\end{aligned}
\right\}. \tag{2.10}
$$

In contrast to all other pure risk models, this formulation includes a constraint in which variables belonging to different scenarios are linked to each other

$$
\sum_{l=1}^{L} \pi_l u_l \leq 1 - \alpha.
$$

Up to now, this did not cause any problems. However, it becomes an issue when we want to exploit the problem's structure for a tailored algorithm. This is explained in Section 2.2.3.

Pure risk model with α-Conditional Value-at-Risk An equivalent deterministic formulation for the pure risk model using the α-Conditional Value-at-Risk as risk measure can be easily deduced if one recalls the description of α-Conditional Value-at-Risk by Expected Excess

$$
\mathbb{Q}_{CVaR_\alpha}(x) := \min \left\{ \eta + \frac{1}{1-\alpha} \mathbb{Q}_{EE_\eta}(x) \; : \; \eta \in \mathbb{R} \right\}.
$$

The following is obtained by Proposition 2.15.

Proposition 2.17
Assume (1.5) – (1.7), that $z(\omega)$ is discretely distributed with finitely many realizations $z_1, \ldots, z_L$ and probabilities $\pi_1, \ldots, \pi_L$, and let $\alpha \in (0,1)$. Then the stochastic program

$$
\min \left\{ \mathbb{Q}_{CVaR_\alpha}(x) \; : \; x \in X \right\}
$$

can be equivalently restated as

$$
\min \left\{ \eta + \tfrac{1}{1-\alpha} \sum_{l=1}^{L} \pi_l v_l \; : \quad
\begin{aligned}
Wy_l + Tx &= z_l && \forall l \\
c^\top x + q^\top y_l - \eta &\leq v_l && \forall l \\
x \in X, \; y_l \in \mathbb{Z}_+^{\bar{m}} \times \mathbb{R}_+^{m'}, \; v_l &\in \mathbb{R}_+, \; \eta \in \mathbb{R} && \forall l
\end{aligned}
\right\}. \tag{2.11}
$$

Hence, the discretization of the random right hand side's probability distribution yields equivalent, mixed-integer linear formulations for all considered mean-risk models. The dimension of the arising problems mainly depends on the number of realizations of $z(\omega)$. The equivalents can become quite large-scale for instances with many data scenarios and therewith intractable for standard solvers. Therefore, a tailored algorithmic approach to solve them is addressed in the following section.

2.2.3 Algorithmic issues – dual decomposition method

Since deterministic equivalents of mean-risk models grow substantially with an increasing number of scenarios, they usually can hardly be solved with standard software for mixed-integer linear programs. In this section we take a closer look at the structure of the equivalents and review a tailored algorithm based on dual decomposition according to the scenarios. Such an algorithm is able to handle the equivalents much more efficiently, which will be shown in Section 4.3 of this thesis, where computational results for the optimal operation of a dispersed generation system gained from mean-risk optimization are presented.

For structural analysis of the deterministic equivalents introduced above lets go back to the equivalent of the pure expectation-based problem (2.7)

$$\min \left\{ c^\top x + \sum_{l=1}^{L} \pi_l q^\top y_l \ : \quad Tx + Wy_l \ = \ z_l \quad \forall l \atop x \in X, \ y_l \in \mathbb{Z}_+^{\bar{m}} \times \mathbb{R}_+^{m'} \quad \forall l \right\}.$$

We first identify the parts of the problem that prevent it from directly decomposing into scenario-specific subproblems. In formulation (2.7) there is no explicit coupling of different scenarios since there are no constraints involving second-stage variables $y_{l'}, y_{l''}$ that belong to different scenarios $l', l'' \in \{1, \dots, L\}$, $l' \neq l''$. But the scenarios are implicitly interlinked due to nonanticipativity of the first-stage variable x which has to be chosen equally for all scenarios. The resulting L-shaped structure of the constraint matrix of problem (2.7) is depicted in Figure 2.1. There T and W are the matrices already used in formulation (2.7), and A shall be a matrix describing all constraints only referring to x.

We introduce copies $x_1, \dots, x_L$ of x and add the explicit nonanticipativity constraints $x_1 = \dots = x_L$. This leads to the reformulation of (2.7)

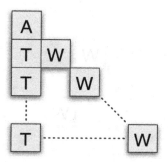

Figure 2.1: Block-structure of the deterministic equivalents for mean-risk models

$$\min \Big\{ \sum_{l=1}^{L} \pi_l (c^\top x_l + q^\top y_l) \; : \quad Tx_l + Wy_l \;\; = \;\; z_l \qquad \forall l$$
$$x_l \in X, \; y_l \in \mathbb{Z}_+^{\bar{m}} \times \mathbb{R}_+^{m'} \quad \forall l \Big\} . \qquad (2.12)$$
$$x_1 = x_2 = \ldots = x_L$$

Obviously, the constraints $Tx_l + Wy_l = z_l$ can be grouped into L-many blocks corresponding to the scenarios that can readily be considered separately. Moreover, the objective function is separable according to the scenarios. Only the explicit nonanticipativity constraints remain which do not allow for the direct decomposition of (2.12). The block-structure of the constraint matrix is visualized in Figure 2.2.

The decomposition of (2.12) into scenario many subproblems can be achieved by Lagrangian relaxation of the explicit nonanticipativity constraints. Moreover, we obtain lower bounds for (2.12), and therewith for (2.7), by the solution of the corresponding Lagrangian dual. Together with heuristics that generate feasible points for (2.7) the lower bounds are finally embedded into a branch-and-bound procedure. Further details of this algorithmic concept are given subsequently.

Note that not only the deterministic equivalent of the expectation-based problem (2.7) shows the structure depicted in Figure 2.2. Almost all equivalents of the pure risk models introduced in Section 2.2.2 maintain these structural properties (see (2.8), (2.9), and (2.11)). Only for the α-Value-at-Risk the latter does not hold. There, additional scenario-coupling enters the deterministic equivalent (2.10) via

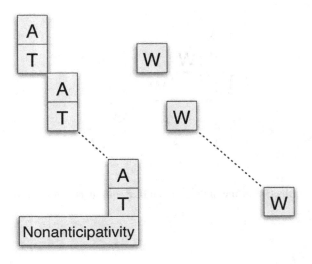

Figure 2.2: Block-structure of the deterministic equivalents for mean-risk models with explicit nonanticipativity

the constraints

$$\sum_{l=1}^{L} \pi_l u_l \leq 1 - \alpha.$$

Of course, there are possibilities to deal with that, but we restrict ourselves to using Excess Probability, Expected Excess and α-Conditional Value-at-Risk as risk measures from now on.

In the following paragraph, we give a detailed description of a dual decomposition procedure yielding lower bounds for the mean-risk problem (2.2). We deduce this procedure exemplarily for the pure expectation-based problem. A second paragraph will show, how to find a feasible point for (2.2) using the solutions gained from lower bounds. Finally, we will explain a branch-and-bound method for solving (2.2), which is well known in mixed-integer programming, and present an algorithm that embeds the lower and upper bounding procedures into the branch-and-bound scheme.

Relaxations and lower bounds   In general, lower bounds for an optimization problem can, for instance, be derived from relaxing some of its constraints and solving the resulting problem. Obviously, that makes the problem "easier" to solve as it enlarges the feasibility set. In case of minimization problems, the optimal solution of a relaxation is smaller or equal to the solution of the original problem. One prominent example of a relaxation for mixed-integer linear problems is the *LP-Relaxation*. It is obtained from the original problem by ignoring integrality requirements to variables.

As mentioned above, we apply *Lagrangian relaxation* to the nonanticipativity constraints to derive a relaxation. This is particularly suitable here since it allows for the scenario-wise decomposition of the problem. For the pure expectation-based problem (2.7) we explain this in detail.

We start with a reformulation of the explicit nonanticipativity constraints $x_1 = x_2 = \ldots = x_L$ as

$$\sum_{l=1}^{L} H_l x_l = 0,$$

where $H_l \in \mathbb{R}^{l' \times m}$, $l = 1, \ldots, L$ (e.g., $l' = m \cdot (L-1)$), are suitable matrices.

This results in the following *Lagrangian function*

$$L(x, y, \lambda) := \sum_{l=1}^{L} \pi_l L_l(x_l, y_l, \lambda),$$

where $\lambda \in \mathbb{R}^{l'}$ and

$$L_l(x_l, y_l, \lambda) := c^\top x_l + q^\top y_l + \lambda^\top H_l x_l.$$

For an arbitrary λ the minimum of the Lagrangian function with respect to the remaining constraints in (2.12) provides a lower bound for problem (2.7). To calculate the best possible lower bound, which can be achieved by this relaxation, we solve the *Lagrangian dual*

$$\max \left\{ D(\lambda) : \lambda \in \mathbb{R}^{l'} \right\}, \tag{2.13}$$

where

$$D(\lambda) = \min \left\{ \sum_{l=1}^{L} \pi_l L_l(x_l, y_l, \lambda) : \begin{array}{ll} Tx_l + Wy_l = z_l & \forall l \\ x_l \in X, \ y_l \in \mathbb{Z}_+^{\bar{m}} \times \mathbb{R}_+^{m'} & \forall l \end{array} \right\}.$$

Due to the structural arguments given above, this optimization problem is separable with respect to individual scenarios

$$D(\lambda) = \sum_{l=1}^{L} \pi_l D_l(\lambda),$$

where, for $l = 1, \ldots, L$,

$$D_l(\lambda) = \min \left\{ L_l(x_l, y_l, \lambda) \; : \; \begin{array}{rcl} Tx_l + Wy_l & = & z_l \\ x_l \in X, \;\; y_l \in \mathbb{Z}_+^{\bar{m}} \times \mathbb{R}_+^{m'} \end{array} \right\}. \tag{2.14}$$

The Lagrangian dual (2.13) is a non-smooth concave maximization (or convex minimization) problem with piecewise linear objective, which can be solved applying advanced bundle methods (see, for instance, [HK02, Kiw90]). Solving the dual, or in other words finding a lower bound then reduces to the computation of a function value and subgradient for $-D(\lambda)$ (when adopting a convex minimization setting in (2.13)). A subgradient of $-D$ at λ is given by

$$-\left(\sum_{l=1}^{L} H_l x_l^{\lambda} \right),$$

where x_l^{λ} are the corresponding components in an optimal solution vector of problem (2.14) defining $D_l(\lambda)$. As a consequence, lower bounds for the original problem (2.7) can be computed by solving the single-scenario mixed-integer linear programs in (2.14) instead of working with the full-size model (2.7). As already mentioned, this decomposition is instrumental since (2.14) may be tractable for standard mixed-integer linear programming solvers while (2.7) may not. For further details see [CS99].

As the pure risk models with Excess Probability, Expected Excess and α-Conditional Value-at-Risk maintain the block-structure of the pure expectation-based problem, lower bounds can be computed by scenario-wise decomposition for these models, too. The above procedure can readily be transferred. Moreover, this means that lower bounds for the mean-risk model (2.2), which arises as a combination of a pure risk model and the expectation-based problem, can be found applying the dual decomposition procedure.

It is known from integer programming that in general there remains an optimality gap when using Lagrangian relaxation (see [NW88]). This means that only $z_{obj}^{LD} \leq z_{obj}^{P}$ holds for the optimal objective value z_{obj}^{LD} of the Lagrangian dual if z_{obj}^{P} denotes the optimal objective value of (2.2). Hence, the optimal solution of the Lagrangian dual usually does not provide a feasible solution of (2.2). Therefore, we have to apply heuristics to derive a feasible solution, which is the topic of the following paragraph.

Upper bounds The objective value $z_{obj}(\bar{x})$ of each feasible solution $\bar{x}$ of problem (2.2) is an upper bound for the optimal solution value z_{obj}^P of (2.2) i.e., it holds that $z_{obj}(\bar{x}) \geq z_{obj}^P$.

If we apply the above described lower bounding procedure to (2.2), we obtain scenario many optimal solutions $\tilde{x}_1, \tilde{x}_2, \ldots, \tilde{x}_L$ from solving the Lagrangian dual. To derive a feasible solution of (2.2) we use them as a starting point. If these proposals for a first-stage decision $\bar{x}$ fulfill the nonanticipativity constraints $x_1 = \ldots = x_L$, we already have found a feasible solution and are done. If the nonanticipativity is not fulfilled, we apply heuristics to the proposals $\tilde{x}_1, \tilde{x}_2, \ldots, \tilde{x}_L$ to recover it. There are several possibilities for such heuristics: For example, the most frequently occurring candidate among the $\tilde{x}_1, \tilde{x}_2, \ldots, \tilde{x}_L$ can be used for $\bar{x}$. Or we calculate the mean value of the $\tilde{x}_1, \tilde{x}_2, \ldots, \tilde{x}_L$, round its components to the next integer if required, and set $\bar{x}$ to this rounded mean value. The best choice of a heuristic depends on the concrete problem we are faced with.

A branch-and-bound algorithm The idea behind the *branch-and-bound procedure* is to partition the feasible set X of problem (2.2) with increasing granularity. This is done by additional linear inequalities which maintain the mixed-integer linear description of the problem. For each subproblem of the partition upper and lower bounds for the optimal objective function value are computed. In our case, we use the above described procedures for these computations. The iterative calculation of lower and upper bounds is embedded into a coordination procedure to guide the partitioning and to prune elements of the partitioning.

Pruning elements of the partitioning can be due to infeasibility, optimality, or inferiority of the subproblems: If a subproblem is infeasible, it is cut of because a further partitioning narrows the feasible set and therefore cannot lead to feasibility again. If a subproblem provides an optimal solution that is feasible for the original problem (2.2) as well, this subproblem has already been solved to optimality which means that further partitioning would not yield an improvement. Finally, if the lower bound of a subproblem is greater than the objective value derived from the currently best solution to (2.2), the subproblem is cut off because there is no chance to remedy the inferiority of the subproblem by further partitioning.

Let $\mathbf{P}$ denote a list of problems and $\varphi_{LB}(P)$ be a lower bound for the optimal value of a subproblem $P \in \mathbf{P}$. By $\bar{\varphi}$ we denote the currently best upper bound to the optimal value of (2.2). The following is a generic description of the branch-and-bound procedure.

Algorithm 2.18

STEP 1 (INITIALIZATION)
Let $\mathbf{P} := \{(2.2)\}$ and $\bar{\varphi} := +\infty$.

STEP 2 (TERMINATION)
If $\mathbf{P} = \emptyset$, then the $\bar{x}$ that yielded $\bar{\varphi} = \mathbb{Q}_{MR}(\bar{x})$ is optimal.

STEP 3 (BOUNDING)
Select and delete a problem P from $\mathbf{P}$. Compute a lower bound $\varphi_{LB}(P)$ and find a feasible solution $\bar{x}$ for P.

STEP 4 (PRUNING)
If $\varphi_{LB}(P) = +\infty$ (infeasibility of the subproblem)
or
$\varphi_{LB}(P) > \bar{\varphi}$ (inferiority of the subproblem), then go to Step 2.

If $\mathbb{Q}_{MR}(\bar{x}) < \bar{\varphi}$, then set $\bar{\varphi} := \mathbb{Q}_{MR}(\bar{x})$.

If $\varphi_{LB}(P) = \mathbb{Q}_{MR}(\bar{x})$ (optimality of the subproblem), then go to Step 2.

STEP 5 (BRANCHING / PARTITIONING)
Create two new subproblems by partitioning the feasible set of P. Add these subproblems to $\mathbf{P}$ and go to Step 2.

If no feasible solution $\bar{x}$ of problem (2.2) was found in Step 3, none of the criteria in Step 4 applies and the algorithm goes to Step 5.

The partitioning in Step 5, in principle, can be carried out by adding (in a proper way) arbitrary linear constraints. The most popular way though is to branch along coordinates. This means to pick a component $x_{(k)}$ of x and add the inequalities $x_{(k)} \leq a$ and $x_{(k)} \geq a+1$ with some integer a if $x_{(k)}$ is an integer component, or, otherwise, add $x_{(k)} \leq a - \varepsilon$ and $x_{(k)} \geq a + \varepsilon$ with some real number a, where $\varepsilon > 0$ is some tolerance parameter to avoid endless branching.

When passing the lower bound $\varphi_{LB}(P)$ of P to its two "children" created in Step 5, the difference, or gap, $\bar{\varphi} - \min_{P \in \mathbf{P}} \varphi_{LB}(P)$ provides information of the quality of the currently best solution. Step 2 often is modified by terminating in case the relative size of this gap drops below some threshold.

3 Stochastic Dominance Constraints induced by Mixed-Integer Linear Recourse

3.1 Introduction to Stochastic Dominance

As described in Section 1.1.1, solving a two-stage stochastic optimization problem corresponds to the selection of a random variable. Applying a traditional mean-risk approach to this problem, requires to select the random variable with respect to statistical parameters reflecting mean and/or risk, like the expectation, the random variable's excess over some preselected ruin level, or the variable's quantiles. In all cases, a "best" random variable is chosen among all possible random variables. Only the definition of "best" distinguishes the different mean-risk formulations.

A new approach is addressed in this chapter that replaces the selection of a "best" random variable by the selection of an acceptable one – according to some benchmark – which simultaneously minimizes a new objective. Therefore, an ordering relation among random variables is defined which yields a ranking and allows to choose only acceptable random variables in the sense of this order. The relation among random variables we use to construct such a ranking is the so-called *stochastic dominance relation*.

The idea of ordering random variables or distribution functions, respectively, is not new. First theorems using orderings similar to stochastic dominance can be found in the work of J. Karamata, published in 1932 ([Kar32]), and G. H. Hardy, J. E. Littlewood, and G. Polya, published in 1959 ([HLP59]). Further developments of these ideas can be found in [Leh55] and [She51]. In the 50's D. Blackwell used similar concepts in the comparison of statistical experiments (see [Bla51, Bla53]).

The application of stochastic dominance to decision theory was introduced in the 50's and 60's. Some examples can be found in [All53, Fis64, Fis70, QS62].

In 1969/70 the first fundamental articles which deal with the extension of stochastic dominance to financial and economical issues were published. For instance, see the work of J. Hadar and W. R. Russell ([HR69, HR71]), G. Hanoch and H. Levy ([HL69]), M. Rothschild and J. E. Stiglitz ([RS70]), and G. A. Whitmore

([Whi70]). Moreover, [MS91] and [WF78] describe the application of stochastic dominance to portfolio selection. Furthermore, articles by V. S. Bawa ([Baw82]) and H. Levy ([Lev92]) can serve as an overview of the publications about stochastic dominance before 1992.

For a modern view on stochastic dominance and other relations comparing random outcomes we refer to the monograph [MS02].

The application of stochastic dominance to stochastic programming as a criterion to select acceptable members from a family of random variables was pioneered by D. Dentcheva and A. Ruszczyński ([DR03a, DR03b, DR03c, DR04a, DR04b, DR06]). For an illustration the models using stochastic dominance constraints are applied to the well-known problem of optimal portfolio selection (see [Mar52, Mar59, Mar87]). To motivate the use of models including stochastic dominance, W. Ogryczak and A. Ruszczyński analyzed the consistency of mean-risk models and stochastic dominance (see e.g., [OR99, OR01, OR02]). The work of D. Dentcheva and A. Ruszczyński focusses on dominance constraints for general random variables enjoying suitable continuity, smoothness, or linearity properties in x and/or z. Under these assumptions results on structure and stability, relaxations, and solution methods for stochastic optimization problems including dominance constraints with discrete probability distributions were already published (see e.g., [DHR07, DR03c, NRR06, NR08]).

In contrast, we establish the theory of first order stochastic dominance constraints induced by mixed-integer linear recourse. Due to the discontinuity of the second-stage value function $\Phi : \mathbb{R}^s \to \mathbb{R}$ this cannot be handled as a special case of the above mentioned work. The well-posedness, structural and stability properties, as well as algorithmic issues shall be analyzed in this chapter. We begin with an introduction to the theory of stochastic dominance relations.

Since first and second order stochastic dominance usually are defined for the preference of maximizing outcomes (profits), we start out from this point of view. Afterwards, we present corresponding results for the situation, where one aims at minimizing losses.

3.1.1 Stochastic orders for the preference of higher outcomes

In this section we assume a framework of maximizing outcomes. In the simple case, where a decision maker can choose between two alternatives with certain prospects, he chooses the one which leads to higher profits. When he has to choose between two actions with uncertain consequences, a decision maker will choose the one promising "in some sense" higher prospects, too. This shall be considered here in detail. We assume that every eligible action and its consequences can be reflected by a random variable and its outcomes. Hence, we aim at a ranking of random variables reflecting a decision maker's preferences.

Such a ranking can be achieved by stochastic order relations, which form partial orders among random variables. In general, a binary relation $\preceq$ on an arbitrary set S is called a *partial order* if it satisfies the following conditions (see [MS02]).

1) *Reflexivity*: $x \preceq x$ for all $x \in S$.

2) *Transitivity*: if $x \preceq y$ and $y \preceq z$ then $x \preceq z$.

3) *Antisymmetry*: if $x \preceq y$ and $y \preceq x$ then $x = y$.

The stochastic order relations presented in this section, especially the stochastic dominance, can be applied to conclude the decisions of rational, risk-averse decision makers who aim at maximizing profits. In the next section we define orders reflecting the preference of minimizing losses.

We presume that all considered random variables have a finite mean i.e., for each random variable X on some probability space $(\Omega, \mathscr{A}, \mathbb{P})$ it holds that

$$\int_\Omega X \, d\mathbb{P} < +\infty.$$

Moreover, by F_X we denote the cumulative distribution function of X which is defined by

$$F_X(t) := \mathbb{P}(X \leq t).$$

Usual Stochastic Order The most natural and prominent candidate for a stochastic order relation might be the *usual stochastic order*, which coincides with first order stochastic dominance defined below. It pointwisely compares the cumulative distribution functions of random variables.

Definition 3.1
A random variable X is said to be smaller than a random variable Y with respect to usual stochastic order, which we write as $X \preceq_{\mathrm{st}} Y$, if

$$F_X(t) \geq F_Y(t) \quad \forall t \in \mathbb{R}.$$

In words, a random variable is smaller with respect to usual stochastic order than another if it adopts smaller values with higher probability.

As the following proposition shows, the usual stochastic order is closely related to the pointwise comparison of random variables.

Proposition 3.2

For random variables X, Y on some probability space $(\Omega, \mathscr{A}, \mathbb{P})$ the following two statements are equivalent:

(i) $X \preceq_{\text{st}} Y$;

(ii) there exist random variables $\hat{X}$ and $\hat{Y}$ on a probability space $(\hat{\Omega}, \hat{\mathscr{A}}, \hat{\mathbb{P}})$ with the cumulative distribution functions F_X and F_Y such that

$$\hat{X}(\omega) \leq \hat{Y}(\omega) \quad \forall \omega \in \Omega.$$

For a proof see [MS02].

First order stochastic dominance – rational decisions   The concept of stochastic dominance can be motivated by the so-called *expected utility hypothesis* by J. von Neumann and O. Morgenstern, which is deduced from some axioms on a rational decision maker in their book "Theory of Games and Economic Behaviour" ([vNM47]). This hypothesis says that every rational decision maker has a *utility function u* such that he prefers an alternative Y to another alternative X if and only if

$$u(X) \leq u(Y).$$

If one knew this utility function, predictions about the choice a decision maker will make (or should make) could be concluded. In fact, usually this utility function is not known exactly, but it is possible to classify utility functions according to suitable assumptions on the decision maker's behavior, which then allows for conclusions about a utility function's characteristics.

As already mentioned, we assume that every rational decision maker will prefer a guaranteed higher outcome to a smaller one. This means that the utility function of a rational decision maker is at least nondecreasing.

Furthermore, if a decision maker is risk-averse, which means that he prefers certain outcomes to risky ones, his utility function u additionally satisfies the inequality

$$\mathbb{E}\big(u(X)\big) \leq u\big(\mathbb{E}(X)\big)$$

for an arbitrary random variable X. Applying Jensen's inequality, leads to the conclusion that u has to be concave.

The first order and second order dominance relation can be characterized using these two classes of functions. Let us start with the definition of the *first order stochastic dominance*:

Definition 3.3
Let X and Y be random variables on some probability space $(\Omega, \mathscr{A}, \mathbb{P})$. Then X is said to dominate Y to first order (in case of the preference of higher outcomes), which we write as $X \succeq_{(1_{\text{big}})} Y$, if

$$\int_\Omega u(X)\, d\mathbb{P} \geq \int_\Omega u(Y)\, d\mathbb{P}$$

holds for all nondecreasing functions u for which these integrals exist.

If we compare two actions and their prospects by first order stochastic dominance relation, this reflects exactly the preferences of a rational decision maker. Therefore, we do not need the complete information about a decision maker's utility function to pick an action he would prefer out of some alternatives.

From the definition we can conclude that

$$X \succeq_{(1_{\text{big}})} Y \Rightarrow \mathbb{E}(X) \geq \mathbb{E}(Y).$$

This can simply be seen if we suppose u to be the identity. It corresponds to the assumption that a rational decider would never choose an action with a minor expected outcome.

Using Proposition 3.2, we can show that first order stochastic dominance can be expressed by usual stochastic order. However, the notion of stochastic dominance is much more common in economics and decision making under risk.

Proposition 3.4
For random variables X, Y it holds

$$X \succeq_{(1_{\text{big}})} Y \Leftrightarrow Y \preceq_{\text{st}} X.$$

A proof can be found in [MS02].

Second order stochastic dominance – risk-averse decisions The *second order stochastic dominance* relation reflects the preferences of a rational risk-averse decider. It is characterized by the class of all concave nondecreasing functions.

Definition 3.5
Let X and Y be random variables on some probability space $(\Omega, \mathscr{A}, \mathbb{P})$. Then X is said to dominate Y to second order (in case of the preference of higher outcomes), which we write as $X \succeq_{(2_{\text{big}})} Y$, if

$$\int_\Omega u(X)\, d\mathbb{P} \geq \int_\Omega u(Y)\, d\mathbb{P}$$

holds for all concave nondecreasing functions u for which these integrals exist.

Obviously, second order stochastic dominance is implied by first order stochastic dominance

$$X \succeq_{(1_{\text{big}})} Y \Rightarrow X \succeq_{(2_{\text{big}})} Y.$$

3.1.2 Stochastic orders for the preference of smaller outcomes

We now transfer first and second order stochastic dominance to a framework, where a decision maker prefers actions which promise lower outcomes to actions promising higher ones. If we think of minimization of losses instead of maximization of profits, this is a reasonable assumption.

First order stochastic dominance – rational decisions   For the preference of smaller outcomes the first order stochastic dominance relation is defined as follows:

Definition 3.6
Let X and Y be random variables on some probability space $(\Omega, \mathscr{A}, \mathbb{P})$. Then X is said to dominate Y to first order (in case of the preference of smaller outcomes), which we write as $X \succeq_{(1_{\text{small}})} Y$, if

$$-X \succeq_{(1_{\text{big}})} -Y.$$

Since we assume X and Y to represent losses, $-X$ and $-Y$ represent profits again. This makes the above definition plausible. We now establish some equivalent characterizations for the first order dominance for the preference of smaller outcomes. We start with a lemma.

Lemma 3.7
For two random variables X and Y the following statements are equivalent:

(i) $X \succeq_{(1_{\text{small}})} Y$;

(ii) $Y \succeq_{(1_{\text{big}})} X$.

Proof
From the definition of first order stochastic dominance for the preference of smaller outcomes we obtain

$$X \succeq_{(1_{\text{small}})} Y \Leftrightarrow -X \succeq_{(1_{\text{big}})} -Y.$$

Together with Definition 3.1 and Proposition 3.4 this yields

$$X \succeq_{(1_{\text{small}})} Y \quad \Leftrightarrow \quad -X \succeq_{(1_{\text{big}})} -Y$$

$$\Leftrightarrow \quad \mathbb{P}(-X \leq \eta) \quad \leq \quad \mathbb{P}(-Y \leq \eta) \qquad \forall \eta \in \mathbb{R}$$

$$\Leftrightarrow \quad \mathbb{P}(X \geq -\eta) \quad \leq \quad \mathbb{P}(Y \geq -\eta) \qquad \forall \eta \in \mathbb{R}$$

$$\Leftrightarrow \quad 1 - \mathbb{P}(X < -\eta) \quad \leq \quad 1 - \mathbb{P}(Y < -\eta) \quad \forall \eta \in \mathbb{R}$$

$$\Leftrightarrow \quad \mathbb{P}(Y < -\eta) \quad \leq \quad \mathbb{P}(X < -\eta) \qquad \forall \eta \in \mathbb{R}$$

$$\Leftrightarrow \quad \mathbb{P}(Y \leq -\eta) \quad \leq \quad \mathbb{P}(X \leq -\eta) \qquad \forall \eta \in \mathbb{R}$$

$$\Leftrightarrow \quad \mathbb{P}(Y \leq \eta) \quad \leq \quad \mathbb{P}(X \leq \eta) \qquad \forall \eta \in \mathbb{R}$$

$$\Leftrightarrow \quad Y \succeq_{(1_{\text{big}})} X.$$

$\square$

We immediately obtain a characterization according to a class of functions.

Proposition 3.8
Let X and Y be random variables on some probability space $(\Omega, \mathscr{A}, \mathbb{P})$. Then $X \succeq_{(1_{\text{small}})} Y$ if and only if

$$\int_{\Omega} u(X) \, d\mathbb{P} \leq \int_{\Omega} u(Y) \, d\mathbb{P}$$

holds for all nondecreasing functions u for which these integrals exist.

This shows that first order dominance for the preference of smaller outcomes coincides with the preference relation of a rational decision maker, who aims at minimal losses. Hence, if an action dominates another one to first order, a rational decision maker prefers this action. Proposition 3.8 can obviously be restated as follows:

Proposition 3.9
Let X and Y be random variables on some probability space $(\Omega, \mathscr{A}, \mathbb{P})$. Then $X \succeq_{(1_{\text{small}})} Y$ if and only if

$$\mathbb{E}\big(u(X)\big) \leq \mathbb{E}\big(u(Y)\big)$$

holds for all nondecreasing functions u for which these expectations exist.

Furthermore, the first order stochastic dominance relation for the preference of smaller outcomes can be characterized by usual stochastic order.

Proposition 3.10

For two random variables X, Y the following statements are equivalent:

(i) $X \succeq_{(1_{\text{small}})} Y$;

(ii) $X \preceq_{\text{st}} Y$;

(iii) $F_X(t) \geq F_Y(t) \ \forall t \in \mathbb{R}$.

Proof

Follows from Lemma 3.7 together with Proposition 3.4. $\square$

Second order stochastic dominance – risk-averse decisions   The second order stochastic dominance for the preference of smaller outcomes is also defined by means of the dominance for the preference of higher outcomes.

Definition 3.11

Let X and Y be random variables on some probability space $(\Omega, \mathscr{A}, \mathbb{P})$. Then X is said to dominate Y to second order (in case of the preference of smaller outcomes), which we write as $X \succeq_{(2_{\text{small}})} Y$, if

$$-X \succeq_{(2_{\text{big}})} -Y.$$

It reflects the preference of a rational risk-averse decider since it can be characterized by the class of all convex nondecreasing functions.

Proposition 3.12

Let X and Y be random variables on some probability space $(\Omega, \mathscr{A}, \mathbb{P})$. Then $X \succeq_{(2_{\text{small}})} Y$ if and only if

$$\int_\Omega u(X)\, d\mathbb{P} \leq \int_\Omega u(Y)\, d\mathbb{P}$$

holds for all convex nondecreasing functions u for which these integrals exist.

Proof

From Definitions 3.5 and 3.11 we obtain

$$X \succeq_{(2_{\text{small}})} Y \quad \Leftrightarrow \quad -X \succeq_{(2_{\text{big}})} -Y$$

$$\Leftrightarrow \quad \int_\Omega u(-X)\, d\mathbb{P} \geq \int_\Omega u(-Y)\, d\mathbb{P} \ \forall \text{ concave nondecreasing functions } u.$$

For $f(x)$ concave it holds that $f(-x)$ is concave and $-f(-x)$ is convex. Moreover, if $f(x)$

is nondecreasing, then $-f(-x)$ is again nondecreasing. This yields

$$\int_\Omega u(-X)\,d\mathbb{P} \;\geq\; \int_\Omega u(-Y)\,d\mathbb{P} \qquad \forall \text{ concave nondecreasing functions } u$$

$$\Leftrightarrow \quad \int_\Omega -u(-Y)\,d\mathbb{P} \;\geq\; \int_\Omega -u(-X)\,d\mathbb{P} \qquad \forall \text{ concave nondecreasing functions } u$$

$$\Leftrightarrow \quad \int_\Omega v(Y)\,d\mathbb{P} \;\geq\; \int_\Omega v(X)\,d\mathbb{P} \qquad \forall \text{ convex nondecreasing functions } v.$$

□

Additionally, we give a more intuitive description of second order stochastic dominance for the preference of smaller outcomes.

Proposition 3.13
A random variable X dominates another random variable Y to second order ($X \succeq_{(2_{\text{small}})} Y$) if and only if

$$\int_\Omega \max\{X-t,0\}\,d\mathbb{P} \leq \int_\Omega \max\{Y-t,0\}\,d\mathbb{P}$$

holds for all $t \in \mathbb{R}$.

Proof
According to Proposition 3.12, it is sufficient to prove that

$$\int_\Omega u(X)\,d\mathbb{P} \leq \int_\Omega u(Y)\,d\mathbb{P} \qquad \forall \text{ convex nondecreasing functions } u$$

$$\Leftrightarrow \quad \int_\Omega \max\{X-t,0\}\,d\mathbb{P} \leq \int_\Omega \max\{Y-t,0\}\,d\mathbb{P} \quad \forall t \in \mathbb{R}.$$

1. Since $u(x) := \max\{x-t,0\}$ is convex and nondecreasing for every $t \in \mathbb{R}$, the second integral inequality obviously follows from the first.

2. Let $u : \mathbb{R} \to \mathbb{R}$ an arbitrary convex and nondecreasing function. Distinguish three cases:

   a) Assume $\lim\limits_{x \to \infty} u(x) = 0$. Then u is the pointwise maximum of a countable set $\{l_1, l_2, \dots\}$ of increasing linear functions. Now define

   $$u_n(x) := \max\{0, l_1(x), l_2(x), \dots, l_n(x)\}.$$

   Then u_n converges to u from below, and each function u_n is piecewise linear with a finite number of kinks. Hence, u_n can be written as

   $$u_n(x) = \sum_{i=1}^{n} a_{in}(x - b_{in})_+$$

for some $a_{in} \in \mathbb{R}_+$ and $b_{in} \in \mathbb{R}$. Thus,

$$\int_\Omega (u_n(X)) \, d\mathbb{P} = \sum_{i=1}^n a_{in} \int_\Omega (X - b_{in})_+ \, d\mathbb{P}$$

$$\leq \sum_{i=1}^n a_{in} \int_\Omega (Y - b_{in})_+ \, d\mathbb{P} = \int_\Omega (u_n(Y)) \, d\mathbb{P}.$$

Then the monotone convergence theorem implies

$$\int_\Omega (u(X)) \, d\mathbb{P} \leq \int_\Omega (u(Y)) \, d\mathbb{P}.$$

b) Assume $\lim_{x \to \infty} u(x) = \alpha \in \mathbb{R}$. Then apply case a) to the function $u - \alpha$.

c) Let $\lim_{x \to \infty} u(x) = -\infty$. Then $u_n(x) := \max\{u(x), -n\}$ fulfills the prerequisites of case b) for all n, and u_n converges to u monotonically. Hence, the assertion again follows from the theorem of monotone convergence.

$\square$

This means that in a framework of minimizing losses X is preferred to Y to second order stochastic dominance if the expected excess of X over an arbitrary fixed value t is not greater than the expected excess of Y over t.

Obviously, the second order stochastic dominance follows from the first order stochastic dominance

$$X \succeq_{(1_{\text{small}})} Y \Rightarrow X \succeq_{(2_{\text{small}})} Y.$$

Subsequently, we establish the first order dominance constrained optimization problem in a minimization framework. For convenience, we use the notations $\succeq_{(1)}$ and $\succeq_{(2)}$ instead of $\succeq_{(1_{\text{small}})}$ and $\succeq_{(2_{\text{small}})}$, respectively.

3.2 Stochastic Dominance Constraints induced by Mixed-Integer Linear Recourse

Stochastic dominance relations form partial orders among random variables. Therefore, we can apply them to the two-stage stochastic optimization problem (1.4) which replaces the selection of a "best" random variable (as done with the mean-risk approach) by the selection of an acceptable member of the family (1.9)

$$\left(c^\top x + \Phi(z(\omega) - Tx) \right)_{x \in X}.$$

The arising stochastic optimization problem with first order dominance constraints induced by mixed-integer linear recourse is described in the following section. Furthermore, we establish some results on the stability of this stochastic problem with respect to perturbations of the underlying probability distributions. Like for mean-risk models, these results allow for the development of deterministic equivalents assuming discrete probability distributions. Finally, a tailored algorithm for such deterministic equivalents based on different relaxations and feasibility heuristics embedded into a branch-and-bound scheme is presented.

3.2.1 First order stochastic dominance constraints

The *stochastic optimization problem with first order stochastic dominance constraints induced by mixed-integer linear recourse* is given by

$$\min_{x} \big\{ g(x) \: : \: f(x,z) \succeq_{(1)} a, \, x \in X \big\}. \tag{3.1}$$

It aims at minimizing the objective $g(x)$ with respect to the first-stage decision $x \in X$. $X \subseteq \mathbb{R}^m$ again denotes a nonempty polyhedron possibly involving integer requirements to x. The function $g : \mathbb{R}^m \to \mathbb{R}$ is assumed to be at least lower semicontinuous and $a : \Omega \to \mathbb{R}$ denotes a random variable on $(\Omega, \mathscr{A}, \mathbb{P})$. Moreover, $f(x,z) = c^\top x + \Phi(z(\omega) - Tx)$ is the random variable on $(\Omega, \mathscr{A}, \mathbb{P})$ representing the costs induced by the first-stage decision x and the resulting optimal recourse action y of the two-stage stochastic optimization problem

$$\min_{x} \big\{ c^\top x + \Phi(z(\omega) - Tx) \: : \: x \in X \big\}$$

with value function

$$\Phi(t) = \min \big\{ q^\top y \: : \: Wy = t, \, y \in \mathbb{Z}_+^{\bar{m}} \times \mathbb{R}_+^{m'} \big\},$$

already defined in Section 1.1.1.

The random variable $a(\omega)$ represents a *reference profile* which serves as a benchmark that has to be dominated by $f(x,z)$ to first order. The function g should be interpreted as new objective that can be chosen corresponding to a decision maker's desires.

Before we establish deterministic equivalents for (3.1) and develop a tailored algorithm, we analyze its structural properties. On the one hand we have to prove the well-posedness of the problem and on the other hand we aim at similar qualitative and quantitative stability results as reviewed for mean-risk models in Section 2.2.1.

In the analysis of the mean-risk models the crucial point were the structural properties of the objective function. For the dominance constrained programs we now focus on the analysis of the feasible region

$$C := \{x \in X : f(x,z) \succeq_{(1)} a\}.$$

It is mainly defined by the integral inequality of the first order dominance relation and substantially depends on the probability distributions of $z(\omega)$ and $a(\omega)$.

3.2.2 Results concerning structure and stability

We start with some structural results for C with fixed distributions of $z(\omega)$ as well as $a(\omega)$ and show that the first order dominance constrained problem (3.1) is well-posed and its minimum is attained. Then we analyze the behavior of the constraint set under small perturbations of the distribution of the random variable $z(\omega)$. A remark also concerns perturbations of both distributions of $z(\omega)$ and $a(\omega)$. We obtain the closedness of the set-valued mapping $C(\mu)$, which assigns the corresponding constraint set to a probability measure μ induced by $z(\omega)$. This allows for proving a qualitative stability result.

Structure of the constraint set for fixed distributions   Let $\mathscr{P}(\mathbb{R}^s)$ and $\mathscr{P}(\mathbb{R})$ be the sets of all Borel probability measures on $\mathbb{R}^s$ and $\mathbb{R}$, respectively. Moreover, let $\mu \in \mathscr{P}(\mathbb{R}^s)$ and $v \in \mathscr{P}(\mathbb{R})$ denote the fixed probability measures induced by the random variables $z(\omega)$ and $a(\omega)$. Applying Proposition 3.10 (equivalence of (i) and (iii)), the constraint set C can be restated as

$$
\begin{aligned}
C &= \{x \in X : F_{f(x,z)}(\eta) \geq F_a(\eta) \, \forall \eta \in \mathbb{R}\} \\
&= \{x \in X : \mu[f(x,z) \leq \eta] \geq v[a \leq \eta] \, \forall \eta \in \mathbb{R}\} \\
&= \{x \in X : \mu[f(x,z) > \eta] \leq v[a > \eta] \, \forall \eta \in \mathbb{R}\}, \quad (3.2)
\end{aligned}
$$

where $F_{f(x,z)}$ denotes the cumulative distribution function of $f(x,z)$, $[f(x,z) \leq \eta]$ is a short notation for $\{z \in \mathbb{R}^s : f(x,z) \leq \eta\}$, and $[a \leq \eta]$ stands for $\{a \in \mathbb{R} : a \leq \eta\}$. For $\delta \in \mathbb{R}$ and $x \in \mathbb{R}^m$ define

$$M_\delta(x) := [f(x,z) > \delta] = \{z \in \mathbb{R}^s : f(x,z) > \delta\}.$$

The following lemma states the semicontinuity of the function that assigns to each $x \in \mathbb{R}^m$ the measure $\mu(M_\delta(x))$.

Lemma 3.14
Assume (1.5) – (1.7) then $\mu(M_\delta(.)) : \mathbb{R}^m \to \mathbb{R}$ is real-valued and lower semicontinuous on $\mathbb{R}^m$.

Proof
Due to the lower semicontinuity of Φ, $M_\delta(x)$ is μ-measurable for each $x \in \mathbb{R}^m$. Therefore, $\mu(M_\delta(.))$ is real-valued.

Let $x, x_n \in \mathbb{R}^m$, $n \in \mathbb{N}$, with $x_n \to x$ and $z \in M_\delta(x)$. The lower semicontinuity of Φ implies

$$\liminf_{x_n \to x} f(x_n, z) \geq f(x, z) > \delta.$$

Hence, there exists a $n_0 \in \mathbb{N}$ such that for all $n \geq n_0$ it holds $f(x_n, z) > \delta$ which implies $z \in M_\delta(x_n)$. Since $\liminf_{x_n \to x} M_\delta(x_n)$ is defined as the set of all points belonging to all except finitely many of the $M_\delta(x_n)$, we obtain that $z \in \liminf_{x_n \to x} M_\delta(x_n)$. This leads to

$$M_\delta(x) \subseteq \liminf_{x_n \to x} M_\delta(x_n).$$

The semicontinuity of the probability measure on sequences of sets then implies

$$\mu(M_\delta(x)) \leq \mu\left(\liminf_{x_n \to x} M_\delta(x_n)\right) \leq \liminf_{x_n \to x} \mu(M_\delta(x_n))$$

which is the asserted lower semicontinuity of $\mu(M_\delta(.))$. $\square$

Using this result, we derive the following proposition.

Proposition 3.15
Assume (1.5) – (1.7) then C is closed.

Proof
Let $x, x_n \in \mathbb{R}^m$, $n \in \mathbb{N}$, with $x_n \to x$ and $x_n \in C$, $n \in \mathbb{N}$, which implies $\mu(M_\eta(x_n)) \leq v[a > \eta]$ for all $\eta \in \mathbb{R}$. Due to the closedness of X, it holds $x \in X$. Applying Lemma 3.14, we have

$$\mu(M_\eta(x)) \leq \liminf_{x_n \to x} \mu(M_\eta(x_n)) \leq \liminf_{x_n \to x} v[a > \eta] = v[a > \eta]$$

for all $\eta \in \mathbb{R}$. Hence, $x \in C$. $\square$

This proposition shows the well-posedness of the dominance constrained problem (3.1) for fixed probability distributions of $z(\omega)$ and $a(\omega)$ as well as bounded, nonempty $X \subseteq \mathbb{R}^m$. The constraint set C is then compact and, therefore, the lower semicontinuous function g attains its minimum on C. In the following section we analyze the case, where the distribution of $z(\omega)$ varies.

Structure and Stability of the constraint set under perturbations of the distributions   Now we assume only the probability distribution of the reference outcome $a(\omega)$ to be fixed and consider the multifunction $C : \mathscr{P}(\mathbb{R}^s) \to 2^{\mathbb{R}^m}$ with

$$C(\mu) := \left\{ x \in \mathbb{R}^m : f(x,z) \succeq_{(1)} a, x \in X \right\} \tag{3.3}$$

and $\mu \in \mathscr{P}(\mathbb{R}^s)$ the probability measure induced by $z(\omega)$.

Furthermore, we equip $\mathscr{P}(\mathbb{R}^s)$ with weak convergence of probability measures ([Bil68, Dud89, VdVW96]), which has already been introduced in Section 2.2.1. We can show that under assumptions (1.5) - (1.7) the multifunction $C(\mu)$ is closed.

Proposition 3.16
Assume (1.5) – (1.7). Then C is a closed multifunction on $\mathscr{P}(\mathbb{R}^s)$. This means that for arbitrary $\mu \in \mathscr{P}(\mathbb{R}^s)$ and sequences $\mu_n \in \mathscr{P}(\mathbb{R}^s)$, $x_n \in C(\mu_n)$, $n \in \mathbb{N}$, with $\mu_n \xrightarrow{w} \mu$ and $x_n \to x$ it follows that $x \in C(\mu)$.

Proof
Let $\mu, \mu_n \in \mathscr{P}(\mathbb{R}^s)$, $n \in \mathbb{N}$, and $x_n \in C(\mu_n)$, $n \in \mathbb{N}$, with $\mu_n \xrightarrow{w} \mu$ and $x_n \to x$. Due to $x_n \in C(\mu_n)$, $n \in \mathbb{N}$, and (3.2), it holds

$$\mu_n[f(x_n,z) \leq \eta] \geq v[a \leq \eta] \quad \forall \eta \in \mathbb{R} \tag{3.4}$$

(again $[f(x_n,z) \leq \eta] = \left\{ z \in \mathbb{R}^s : f(x_n,z) \leq \eta \right\}$ and $[a \leq \eta] = \left\{ a \in \mathbb{R} : a \leq \eta \right\}$).

Like before, define $M_\delta(x) := [f(x,z) > \delta]$. Using this notation, we can reformulate (3.4) as

$$\mu_n(M_\eta(x_n)) + v[a \leq \eta] \leq 1 \quad \forall \eta \in \mathbb{R}. \tag{3.5}$$

The assumptions (1.5) – (1.7) yield the lower semicontinuity of Φ and, therewith, of $f(x,.)$. Hence, $M_\delta(x)$ is an open set for all $\delta \in \mathbb{R}$ and $x \in \mathbb{R}^m$, and we can apply the Portmanteau Theorem (see [Bil68], Theorem 2.1.). This implies that

$$\mu(M_\eta(x)) \leq \liminf_{n\to\infty} \mu_n(M_\eta(x)) \quad \forall \eta \in \mathbb{R}. \tag{3.6}$$

Furthermore, the lower semicontinuity of Φ yields

$$M_\eta(x) \subseteq \liminf_{k\to\infty} M_\eta(x_k) \quad \forall \eta \in \mathbb{R}, \tag{3.7}$$

where "lim inf" denotes the set-theoretic limes inferior i.e., the set of all points belonging to all but a finite number of sets $M_\eta(x_k)$.

Considering (3.7) and the lower semicontinuity of the probability measure (see [Bil86], Theorem 4.1), we obtain for a fixed $n \in \mathbb{N}$

$$\mu_n(M_\eta(x)) \leq \mu_n\left(\liminf_{k\to\infty} M_\eta(x_k)\right) \leq \liminf_{k\to\infty} \mu_n(M_\eta(x_k)) \quad \forall \eta \in \mathbb{R}. \tag{3.8}$$

Taking the limes inferior with respect to $n \in \mathbb{N}$ in (3.8), it holds that

$$\liminf_{n \to \infty} \mu_n(M_\eta(x)) \leq \liminf_{n \to \infty} \liminf_{k \to \infty} \mu_n(M_\eta(x_k)) \leq \liminf_{n \to \infty} \mu_n(M_\eta(x_n)) \quad \forall \eta \in \mathbb{R}. \quad (3.9)$$

For the last inequality we pick the diagonal sequence with $k = n$.

From the inequalities (3.9) and (3.6) we now obtain

$$\mu(M_\eta(x)) \leq \liminf_{n \to \infty} \mu_n(M_\eta(x_n)) \quad \forall \eta \in \mathbb{R}. \quad (3.10)$$

Finally, we use (3.10) and (3.5) and arrive at

$$\mu(M_\eta(x)) + v[a \leq \eta] \leq \liminf_{n \to \infty} \mu_n(M_\eta(x_n)) + v[a \leq \eta] \leq 1 \quad \forall \eta \in \mathbb{R}.$$

Together with (3.2) and (3.3) this implies that $f(x,z) \succeq_{(1)} a$. Since $x_n \to x$ and X is closed, we have $x \in X$. Altogether it follows that $x \in C(\mu)$, which means that C is a closed multifunction. $\square$

If $\mathscr{P}(\mathbb{R})$ is equipped with *uniform convergence of cumulative distribution functions* (Kolmogorov-Smirnov convergence), as done in [DHR07] for example, a similar result can be proven for the case, where additionally the probability measure $v \in \mathscr{P}(\mathbb{R})$ induced by the reference outcome $a(\omega)$ is not fixed.

Proposition 3.17

Assume (1.5) – (1.7). Then the multifunction $\bar{C} : \mathscr{P}(\mathbb{R}^s) \times \mathscr{P}(\mathbb{R}) \to 2^{\mathbb{R}^m}$ with

$$\bar{C}(\mu, v) := \left\{ x \in \mathbb{R}^m : f(x,z) \succeq_{(1)} a, x \in X \right\}$$

is closed.

Proof

Let $v_n, v \in \mathscr{P}(\mathbb{R})$, $n \in \mathbb{N}$, and $v_n \to v$ in Kolmogorov-Smirnov sense. Then it holds in particular that

$$v_n[a \leq \eta] \to v[a \leq \eta] \quad \forall \eta \in \mathbb{R}. \quad (3.11)$$

Further, let $\mu, \mu_n \in \mathscr{P}(\mathbb{R}^s)$, $n \in \mathbb{N}$, and $x_n \in \bar{C}(\mu_n, v_n)$, $n \in \mathbb{N}$, with $\mu_n \overset{w}{\longrightarrow} \mu$ and $x_n \to x$. Considering (3.2), $x_n \in \bar{C}(\mu_n, v_n)$, implies

$$\mu_n[f(x_n, z) \leq \eta] \geq v_n[a \leq \eta] \quad \forall \eta \in \mathbb{R}$$

which can be restated as

$$\mu_n(M_\eta(x_n)) + v_n[a \leq \eta] \leq 1. \quad (3.12)$$

Like in the preceding proof we obtain

$$\mu(M_\eta(x)) \leq \liminf_{n \to \infty} \mu_n(M_\eta(x_n)) \quad \forall \eta \in \mathbb{R}. \quad (3.13)$$

Using (3.13), (3.12), and (3.11), we can conclude that

$$\mu(M_\eta(x)) + \nu[a \leq \eta] \leq \liminf_{n \to \infty} \mu_n(M_\eta(x_n)) + \liminf_{n \to \infty} \nu_n[a \leq \eta] \leq 1.$$

Together with the closedness of X this yields $x \in \bar{C}(\mu, \nu)$. $\square$

Remark 3.18
Proposition 3.17 breaks down when equipping $\mathscr{P}(\mathbb{R})$ only with weak convergence of probability measures instead of the stronger convergence in the Kolmogorov-Smirnov sense. To see this, let $\eta = 0$ and $\mu, x, f(x,z)$ be such that $\mu(M_0(x)) = \frac{1}{2}$. Furthermore, let ν_n, ν be the discrete probability measures with mass 1 at $\frac{1}{n}$ and 0, respectively. Then ν_n converges weakly to ν and it holds

$$\mu(M_0(x)) + \nu_n[a \leq 0] = \frac{1}{2} + 0 \leq 1 \quad \forall n$$

which yields $x \in \bar{C}(\mu, \nu_n)$ for all n (see (3.12)). On the other hand $1 = \nu[a \leq 0]$ and, therewith,

$$\mu(M_0(x)) + \nu[a \leq 0] = \frac{1}{2} + 1 \nleq 1.$$

Hence, $x \notin \bar{C}(\mu, \nu)$.

In [DHR07] a similar closedness result for stochastic problems with first order dominance constraints is proven for a different class of random variables. There stronger continuity of the random variables is assumed and the convergence of probability measures is given by a suitable discrepancy on the counterpart of $\mathscr{P}(\mathbb{R}^s)$. In contrast, we allow discontinuities of the supposed random variables and equip $\mathscr{P}(\mathbb{R}^s)$ with weak convergence of probability measures, which is a weaker notion of convergence than the discrepancy in [DHR07].

The closedness result 3.16 for the constraint set mapping can now be used to derive a stability result for the stochastic program (3.1) under small perturbations of the underlying distributions. As done in Section 2.2.1 for the mean-risk models, we consider the dominance constrained problem as a special parametric optimization problem with the probability measure $\mu \in \mathscr{P}(\mathbb{R}^s)$ induced by $z(\omega)$ as varying parameter

$$P(\mu) \qquad \min \left\{ g(x) : x \in C(\mu) \right\}.$$

The qualitative stability result then is the following.

Proposition 3.19
Assume (1.5) – (1.7). Let X be a nonempty and compact set and g at least lower semicontinuous. Suppose $\bar{\mu} \in \mathscr{P}(\mathbb{R}^s)$ such that the parametric problem $P(\bar{\mu})$ has an optimal solution.

Then the optimal value function $\varphi(\mu) := \inf\{g(x) : x \in C(\mu)\}$ is lower semicontinuous at $\bar{\mu}$ i.e., for $\mu_n \in \mathscr{P}(R^s)$, $n \in \mathbb{N}$, with $\mu_n \xrightarrow{w} \bar{\mu}$ it holds that

$$\varphi(\bar{\mu}) \leq \liminf_{n \to \infty} \varphi(\mu_n).$$

Proof

Let $\bar{\mu}, \mu_n \in \mathscr{P}(R^s)$, $n \in \mathbb{N}$, and $\mu_n \xrightarrow{w} \bar{\mu}$. If $C(\mu_n) = \emptyset$ for any n, then it is $\varphi(\mu_n) = +\infty$, which does not violate the inequality $\varphi(\bar{\mu}) \leq \liminf_{n \to \infty} \varphi(\mu_n)$. Hence, assume without loss of generality that $C(\mu_n) \neq \emptyset$ for all n.

Let $\varepsilon \in \mathbb{R}_+$ be arbitrarily fixed. Then there exists a $x_n \in C(\mu_n)$ with

$$g(x_n) \leq \varphi(\mu_n) + \varepsilon. \tag{3.14}$$

Since $x_n \in X$ for all n and X is compact, there exists an accumulation point $\bar{x} \in X$ of the sequence $(x_n)_{n \in \mathbb{N}}$.

The closedness result on $C(\mu)$ (see Proposition 3.16) yields $\bar{x} \in C(\bar{\mu})$. Together with the assumed lower semicontinuity of g and (3.14) this gives us

$$\varphi(\bar{\mu}) \leq g(\bar{x}) \leq \liminf_{n \to \infty} g(x_n) \leq \liminf_{n \to \infty} \varphi(\mu_n) + \varepsilon.$$

With $\varepsilon \to 0$ this implies

$$\varphi(\bar{\mu}) \leq \liminf_{n \to \infty} \varphi(\mu_n)$$

which proves the lower semicontinuity of $\varphi(\mu)$ at $\bar{\mu}$. $\qquad\square$

This qualitative result allows for the approximative solution of the first order dominance constrained optimization problem by appropriate discretization of the probability distribution of $z(\omega)$. It states that each accumulation point of approximative solutions is at least feasible for the exact problem. An appropriate discretization of the distribution of $z(\omega)$ can be derived by approximation schemes. Here, each approximation scheme can be applied that guarantees weak convergence of the approximative probability measure to the exact one. As mentioned in Section 2.2.2, a discretization via almost surely converging densities (Scheffé's Theorem [Bil68]), via conditional expectations (see [BW86, Kal87]), or via estimation using empirical measures (Glivenko-Cantelli almost sure uniform convergence [Pol84]) often lead to weakly converging probability measures, for instance.

Furthermore, the result shows that in the worst case we overestimate the optimal objective value by an approximation. For real world problems this is much less critical than an underestimation would be.

For a quantitative stability result that allows an assertion on the deviation of the approximative optimal objective value from the exact one the closedness result on $C(\mu)$ is not sufficient. First of all, Proposition 3.16 together with the assumption that X is compact gives us the *upper semicontinuity of $C(\mu)$ according to Hauss-dorff* (see [BGK$^+$82]). This means that for every $\mu_0 \in \mathscr{P}(\mathbb{R}^s)$ and $\varepsilon > 0$ there exists a $\delta > 0$ such that

$$C(\mu) \subseteq \mathscr{U}_\varepsilon(C(\mu_0)) \quad \forall \mu \in \mathscr{U}_\delta(\{\mu_0\}).$$

However, for quantitative stability we additionally need the lower semicontinuity of $C(\mu)$ at least according to Berge ([BGK$^+$82]): A set-valued mapping is called *lower semicontinuous according to Berge* if for every $\mu_0 \in \mathscr{P}(\mathbb{R}^s)$ and open set $\mathscr{O}$ satisfying $\mathscr{O} \cap C(\mu_0) \neq \emptyset$ there exists a $\delta > 0$ such that

$$C(\mu) \cap \mathscr{O} \neq \emptyset \quad \forall \mu \in \mathscr{U}_\delta(\{\mu_0\}).$$

One approach to derive this lower semicontinuity and therewith the desired quantitative stability result is based on *metric regularity* of the function determining the constraint set. Details can be found in [HR99], where the stability of stochastic programs with probabilistic constraints is considered.

This approach mainly makes use of the following proposition, relying on [Kla94], which states a quantitative stability result for the chance constrained problem

$$P(\mu) \qquad \min\{g(x) : x \in C, \ \mu(H_j(x)) \geq p_j, \ j = 1,\ldots,d\}$$

with $C \subseteq \mathbb{R}^m$, $\mu \in \mathscr{P}(\mathbb{R}^s)$, a localized optimal value function, and a solution set mapping on a subset $V \subseteq \mathbb{R}^m$

$$\varphi_V(\mu) := \inf\{g(x) : x \in C \cap \mathrm{cl}V, \ \mu(H_j(x)) \geq p_j, \ j = 1,\ldots,d\},$$
$$\psi_V(\mu) := \{x \in C \cap \mathrm{cl}V, \ \mu(H_j(x)) \geq p_j, \ j = 1,\ldots,d : g(x) = \varphi_V(\mu)\}.$$

Proposition 3.20
Assume that X is a CLM set for $P(\mu)$ with respect to a bounded set V (i.e., $X = \psi_V(\mu)$ and X compact), that g is locally Lipschitz continuous, and the probabilistic constraint function $\Theta_\mu(.) - p : \mathbb{R}^m \to \mathbb{R}^d$ with $p = (p_1,\ldots,p_d)$ and

$$\Theta_\mu^j(x) - p_j := \mu(H_j(x)) - p_j, \ j = 1,\ldots,d,$$

is metrically regular with respect to C at each $x^0 \in C$. Then there exist constants $L > 0$ and $\delta > 0$ such that the set valued mapping Ψ_V is upper semicontinuous at μ, $\Psi_V(\nu)$ is a CLM set for $P(\nu)$ with respect to V, and

$$|\varphi_V(\mu) - \varphi_V(\nu)| \leq L \cdot \alpha_{\mathscr{B}}(\mu,\nu)$$

holds whenever $\nu \in \mathscr{P}(\mathbb{R}^s)$ and $\alpha_{\mathscr{B}}(\mu,\nu) < \delta$.

Here $\alpha_{\mathscr{B}}$ denotes a distance, sometimes called $\mathscr{B}$-discrepancy, which for $\mu, \nu \in \mathscr{P}(\mathbb{R}^s)$ is defined as

$$\alpha_{\mathscr{B}}(\mu, \nu) := \sup\big\{\, |\mu(B) - \nu(B)| \, : \, B \in \mathscr{B} \,\big\},$$

where $\mathscr{B}$ is a class of closed subsets of $\mathbb{R}^s$ such that all sets of the form $H_j(x)$ ($x \in C$, $j = 1, \dots, d$) belong to $\mathscr{B}$ and that $\mathscr{B}$ is a determining class.

The proposition can be applied to the first order dominance constrained optimization problem. The crucial point is then to prove the metric regularity of the function defining the constraint set. This can be done via a growth condition (see [HR99]) and should at least be possible if the dominance constraints are induced by linear instead of mixed-integer linear recourse. This will be future work based on results presented in this thesis.

For a more general class of random variables enjoying suitable smoothness, continuity, and linearity conditions such a quantitative result has been proven. Both, qualitative and quantitative results for these first order dominance constrained problems were established in [DHR07].

3.2.3 Deterministic equivalents

The stochastic optimization problem with first order dominance constraints

$$\min_x \big\{\, g(x) \, : \, f(x, z) \succeq_{(1)} a, \; x \in X \,\big\}$$

contains a continuum of constraints involving multivariate integrals

$$f(x, z) \succeq_{(1)} a \;\Leftrightarrow\; \mu[f(x, z) \le \eta] \ge \nu[a \le \eta] \quad \forall \eta \in \mathbb{R}.$$

Therefore, it is not readily tractable by solution algorithms. A discretization of the underlying probability distributions turns out helpful. As already explained, the stability result of the above section provides justification. In a first step we assume the reference profile $a(\omega)$ to be discretely distributed. Then we additionally deal with the discrete distribution of $z(\omega)$ with finitely many scenarios and corresponding probabilities.

Proposition 3.21
Let $a(\omega)$ be discretely distributed with finitely many realizations a_k and probabilities p_k, $k = 1, \dots, D$ and $f(x, z)$ the random costs induced by some fixed x living on the probability space $(\Omega, \mathscr{A}, \mathbb{P})$. Moreover, assume that $\nu \in \mathscr{P}(\mathbb{R})$ and $\mu \in \mathscr{P}(\mathbb{R}^s)$ denote the probability measures induced by the random variables $a(\omega)$ and $z(\omega)$. Then

$$f(x, z) \succeq_{(1)} a$$

is equivalent to

$$\mu[f(x,z) \leq a_k] \geq v[a \leq a_k] \quad \text{for } k = 1, \ldots, D. \tag{3.15}$$

Proof
By Proposition 3.10 we have $f(x,z) \succeq_{(1)} a$ if and only if $\mu[f(x,z) \leq \eta] \geq v[a \leq \eta]$ for all $\eta \in \mathbb{R}$. This obviously implies $\mu[f(x,z) \leq a_k] \geq v[a \leq a_k]$ for all k.

Assume there exists an $a_0 \in \mathbb{R}$ with $v[a \leq a_0] = 0$ and the realizations of $a(\omega)$ are given in an ascending order $a_0 < a_1 < \cdots < a_D$. Consider three cases:

a) Let $\eta \in \mathbb{R}$ with $a_{i-1} \leq \eta < a_i$, $2 \leq i \leq D$. Since there is no mass point of $a(\omega)$ in between a_{i-1} and a_i, it holds

$$v[a \leq \eta] = v[a \leq a_{i-1}], \tag{3.16}$$

and the monotonicity of the cumulative distribution function implies

$$\mu[f(x,z) \leq a_{i-1}] \leq \mu[f(x,z) \leq \eta]. \tag{3.17}$$

The combination of the two inequalities (3.16) and (3.17) using (3.15) gives us

$$v[a \leq \eta] \leq \mu[f(x,z) \leq \eta].$$

b) Let $\eta < a_1$, which yields $v[a \leq \eta] = 0$. Since the cumulative distribution function is nonnegative, it holds $v[a \leq \eta] \leq \mu[f(x,z) \leq \eta]$.

c) Let $\eta \geq a_D$, which yields $v[a \leq \eta] = v[a \leq a_D] = 1$. Together with (3.15), the monotonicity of the cumulative distribution function, and the fact that the cumulative distribution function is not greater than 1 we obtain

$$1 = v[a \leq a_D] \leq \mu[f(x,z) \leq a_D] \leq \mu[f(x,z) \leq \eta] \leq 1.$$

This implies $\mu[f(x,z) \leq \eta] = 1 = v[a \leq \eta]$.

Altogether we arrive at $\mu[f(x,z) \leq \eta] \geq v[a \leq \eta]$ for all $\eta \in \mathbb{R}$. □

If we additionally assume $z(\omega)$ to be discretely distributed, we obtain the following deterministic equivalent of the first order dominance constrained optimization problem.

Proposition 3.22
Assume (1.5) – (1.7). Let $z(\omega)$ and $a(\omega)$ in problem (3.1) be discretely distributed with realizations z_l, $l = 1, \ldots, L$, and a_k, $k = 1, \ldots, D$, as well as probabilities π_l, $l = 1, \ldots, L$, and p_k, $k = 1, \ldots, D$, respectively. Moreover, let $\mu \in \mathscr{P}(\mathbb{R}^s)$ and $v \in \mathscr{P}(\mathbb{R})$ denote the probability measures induced by $z(\omega)$ as well as $a(\omega)$, let $g(x) := g^\top x$ be linear, and let X

be bounded. Then there exists a constant M such that (3.1) is equivalent to the mixed-integer linear program

$$\min\Big\{g^\top x: \quad \begin{aligned} c^\top x + q^\top y_{lk} - a_k &\leq M\theta_{lk} && \forall l\,\forall k \\[2mm] \sum_{l=1}^{L}\pi_l\theta_{lk} &\leq \bar{a}_k && \forall k \\[2mm] Tx + Wy_{lk} &= z_l && \forall l\,\forall k \\[2mm] x\in X,\; y_{lk}\in\mathbb{Z}_+^{\bar{m}}\times\mathbb{R}_+^{m'},\; \theta_{lk}\in\{0,1\} && \forall l\,\forall k \end{aligned}\Big\}, \tag{3.18}$$

where $\bar{a}_k := 1 - \nu[a \leq a_k]$, $k = 1,\ldots,D$.

Proof
We start with the construction of M and show its existence. Let

$$M > \sup\big\{c^\top x + \Phi(z_l - Tx) - a_k : x\in X,\, l=1,\ldots,L,\, k=1,\ldots,D\big\}.$$

For the existence of M we prove that the above supremum is finite. By Proposition 2.1 in Section 2.2.1 we know that there exist positive constants β and γ such that

$$|\Phi(t_1) - \Phi(t_2)| \leq \beta\|t_1 - t_2\| + \gamma$$

for $t_1, t_2 \in \mathbb{R}^s$.

Furthermore, (1.7) implies that $\Phi(0) = 0$. This enables the following estimation for each l and k

$$\begin{aligned} |c^\top x + \Phi(z_l - Tx) - a_k| &\leq |c^\top x| + |\Phi(z_l - Tx) - \Phi(0)| + |a_k| \\[2mm] &\leq \|c\|\|x\| + \beta\|z_l - Tx\| + \gamma + |a_k| \\[2mm] &\leq \|c\|\cdot\|x\| + \beta\|z_l\| + \beta\|T\|\cdot\|x\| + \gamma + |a_k|. \end{aligned}$$

Since X is bounded and there are only finitely many z_l and a_k, the last expression is finite which proves the existence of M.

We continue with a reformulation of problem (3.1). From (3.2) and Propostion 3.21 we obtain that (3.1) is equivalent to

$$\min\big\{g^\top x : \mu[f(x,z) \leq a_k] \geq \nu[a \leq a_k]\ \forall k,\, x\in X\big\}.$$

With $\bar{a}_k := 1 - \nu[a \leq a_k]$, $k = 1,\ldots,D$, this can be restated as

$$\min\big\{g^\top x : \mu[f(x,z) > a_k] \leq \bar{a}_k\ \forall k,\, x\in X\big\}. \tag{3.19}$$

Finally, we prove the proposition by considering the following two sets for any $k\in\{1,\ldots,D\}$

$$S_{1k} := \{x\in X : \mu[f(x,z) > a_k] \leq \bar{a}_k\},$$

$$S_{2k} := \left\{ x \in X : \quad \exists\, y_l \in \mathbb{Z}_+^{\bar{m}} \times \mathbb{R}_+^{m'}, \ \exists\, \theta_l \in \{0,1\} \ \forall l \right.$$

such that :

$$c^\top x + q^\top y_l - a_k \ \leq \ \mathsf{M}\theta_l$$

$$Tx + Wy_l \ = \ z_l$$

$$\sum_{l=1}^{L} \pi_l \theta_l \ \leq \ \bar{a}_k$$

$$\left. \right\}.$$

Having in mind (3.19), it holds that $\bigcap\limits_{k=1}^{D} S_{1k}$ and $\bigcap\limits_{k=1}^{D} S_{2k}$ equal the constraint sets of (3.1) and (3.18), respectively. Hence, it remains to show that $S_{1k} = S_{2k}$ for all $k \in \{1,\dots,D\}$.

1. Let $x \in S_{1k}$, consider $I := \{l \in \{1,\dots,L\} : f(x,z_l) > a_k\}$, and put

$$\theta_l := \begin{cases} 1 & \text{if } l \in I \\[2mm] 0 & \text{else.} \end{cases}$$

This yields

$$\sum_{l=1}^{L} \pi_l \theta_l = \sum_{l \in I} \pi_l \theta_l \leq \bar{a}_k.$$

For $l \notin I$ we have $f(x,z_l) = c^\top x + \Phi(z_l - Tx) \leq a_k$. Hence, there exists a $y_l \in \mathbb{Z}_+^{\bar{m}} \times \mathbb{R}_+^{m'}$ with

$$c^\top x + q^\top y_l - a_k \leq 0 = \mathsf{M}\theta_l$$

and

$$Tx + Wy_l = z_l.$$

For $l \in I$ we have $f(x,z_l) = c^\top x + \Phi(z_l - Tx) > a_k$. Therefore, we can choose $y_l \in \mathbb{Z}_+^{\bar{m}} \times \mathbb{R}_+^{m'}$ such that $Tx + Wy_l = z_l$ and $q^\top y_l = \Phi(z_l - Tx)$. By the construction of M we obtain

$$c^\top x + q^\top y_l - a_k \leq \mathsf{M} = \mathsf{M}\theta_l.$$

Altogether it follows that $x \in S_{2k}$ and therewith $S_{1k} \subseteq S_{2k}$.

2. Let $x \in S_{2k}$ and consider $J := \{l \in \{1,\dots,L\} : \theta_l = 1\}$. As for each $l \notin J$ there exists a $y_l \in \mathbb{Z}_+^{\bar{m}} \times \mathbb{R}_+^{m'}$ such that

$$c^\top x + q^\top y_l - a_k \leq 0$$

and

$$Tx + Wy_l = z_l,$$

we have $f(x,z_l) = c^\top x + \Phi(z_l - Tx) \leq a_k$ for all $l \notin J$. This yields

$$\{l \in \{1,\dots,L\} : f(x,z_l) > a_k\} \subseteq J.$$

We obtain

$$\mu[f(x,z_l) > a_k] \le \sum_{l \in J} \pi_l \theta_l = \sum_{l=1}^{L} \pi_l \theta_l \le \bar{a}_k$$

which implies $x \in S_{1k}$ and therewith $S_{2k} \subseteq S_{1k}$.

Case 1. and 2. together finally yield $S_{1k} = S_{2k}$, which completes the proof. $\qquad\square$

In the above described deterministic equivalent we use one second-stage variable y_{lk} for each scenario $l = 1,\ldots,L$ and each reference profile $k = 1,\ldots,D$, which turns out to be not necessary. As the following proposition states, it is possible to show that the second-stage variables only need to depend on the scenarios but not on the reference profiles, which reduces the number of constraints substantially.

Proposition 3.23
Assume (1.5) – (1.7). Let $z(\omega)$ and $a(\omega)$ in problem (3.1) be discretely distributed with realizations z_l, $l = 1,\ldots,L$, and a_k, $k = 1,\ldots,D$, as well as probabilities π_l, $l = 1,\ldots,L$, and p_k, $k = 1,\ldots,D$, respectively. Moreover, let $\mu \in \mathscr{P}(\mathbb{R}^s)$ and $\nu \in \mathscr{P}(\mathbb{R})$ denote the probability measures induced by $z(\omega)$ and $a(\omega)$, let $g(x) := g^\top x$ be linear, and let X be bounded. Then there exists a constant M such that (3.1) is equivalent to the mixed-integer linear program

$$\min \left\{ g^\top x : \quad
\begin{aligned}
c^\top x + q^\top y_l - a_k &\le M\theta_{lk} & \forall l\, \forall k \\[4pt]
\sum_{l=1}^{L} \pi_l \theta_{lk} &\le \bar{a}_k & \forall k \\[4pt]
Tx + W y_l &= z_l & \forall l \\[4pt]
x \in X,\ y_l \in \mathbb{Z}_+^{\bar{m}} \times \mathbb{R}_+^{m'},\ \theta_{lk} &\in \{0,1\} & \forall l\, \forall k
\end{aligned}
\right\}, \qquad (3.20)$$

where $\bar{a}_k := 1 - \nu[a \le a_k]$, $k = 1,\ldots,D$.

Proof
Taking into account Proposition 3.22, it is sufficient to show the equivalence of (3.18) and (3.20). Obviously, each feasible solution $(\bar{x},\bar{y},\bar{\theta})$ of (3.20) provides a feasible solution of (3.18) with the same objective value if we choose $y_{lk} = \bar{y}_l$ for each $k \in \{1,\ldots,D\}$.

Now assume $(\bar{x},\bar{y},\bar{\theta})$ is a feasible solution of (3.18). For each l choose $\hat{y}_l$ from the set $\{\bar{y}_{lk} : k = 1,\ldots,D\}$ with

$$\hat{y}_l = \arg\min\{q^\top \bar{y}_{lk} : k = 1,\ldots,D\}.$$

It holds

$$c^\top x + q^\top \hat{y}_l - a_k \le c^\top x + q^\top \bar{y}_{lk} - a_k \le M\theta_{lk} \quad \forall l\, \forall k$$

and

$$Tx + W\hat{y}_l = z_l \quad \forall l,$$

which yields the feasibility of $(\bar{x}, \hat{y}, \bar{\theta})$ for (3.20). Since the objective function only depends on the first-stage variable $\bar{x}$, the proof is complete.   □

Though using formulation (3.20) reduces the number of constraints, this formulation is not always preferable for the solution of the deterministic equivalents. In many cases the usage of a sparser constraint matrix, which arises when the second-stage variables are copied corresponding to the reference profiles, as in (3.18), yields a better performance. Some computational tests have been made for the energy optimization problem described in Section 4.3. They prove the superiority of the sparser problem (3.18). The standard mixed-integer solver CPLEX ([ILO05]) solves small instances of (3.18) up to optimality. However, often no feasible solution of (3.20) can be found in a given timelimit of eight hours.

3.2.4 Algorithmic issues

Both described deterministic equivalents are mixed-integer linear programming problems that, in general, can readily be tackled by standard mixed-integer linear programming solvers. However, with an increasing number of scenarios L and reference profiles D the equivalents become very large-scale which carries standard solvers to their limits. To overcome these computational limitations, we develop some special algorithmic methods in this section allowing for the problems' decomposition.

The algorithmic approaches are designed to exploit the special structure of the stochastic problem with first order dominance constraints. As done for the expectation-based problem in Section 2.2.3, we start with an analysis of the problem's constraint matrix and figure out the constraints that interlink different scenarios l. For convenience, we do this for the reduced formulation (3.20), but all considerations can readily be transferred to the extended formulation (3.18), too. Figure 3.1 schematically shows the constraint matrix of (3.20) except for some slack variables' coefficients.

Compared to the structure of the pure expectation-based model (2.7), which was displayed in Figure 2.1, there are many similarities, but also some differences. Again, we have scenario-specific blocks that are implicitly interlinked by the nonanticipativity of the first-stage variable x which occurs in each block, but must be chosen equally in each scenario. The new constraints $c^\top x + q^\top y_l - a_k \leq M\theta_{lk}$ are covered by these blocks as well.

In contrast to the pure expectation-based case, there are additional constraints explicitly involving variables that belong to different scenarios. The block "Dom-

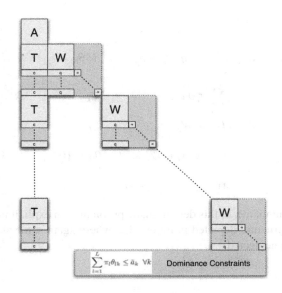

Figure 3.1: Block-structure of the reduced deterministic equivalent for a dominance constrained optimization problem

inance Constraints" represents D inequalities

$$\sum_{l=1}^{L} \pi_l \theta_{lk} \leq \bar{a}_k \ \forall k$$

that control the excess of $f(x,z)$ over the profile $a(\omega)$. Despite this linkage of scenarios, the form of the constraint matrix gives rise to a decomposition approach. As done with the expectation-based model, we introduce copies $x_l, l = 1, \ldots, L$ of the first-stage variable x and replace the implicit nonanticipativity by an explicit formulation

$$\min\Big\{ \sum_{l=1}^{L} \pi_l g^\top x_l : \quad c^\top x_l + q^\top y_l - a_k \;\le\; M\theta_{lk} \qquad \forall l \; \forall k$$

$$\sum_{l=1}^{L} \pi_l \theta_{lk} \qquad\qquad \le\; \bar{a}_k \qquad \forall k$$

$$T x_l + W y_l \qquad\qquad =\; z_l \qquad \forall l \Big\}. \qquad (3.21)$$

$$x_l \in X, \; y_l \in \mathbb{Z}_+^{\bar{m}} \times \mathbb{R}_+^{m'}, \; \theta_{lk} \in \{0,1\} \quad \forall l \; \forall k$$

$$x_1 = x_2 = \ldots = x_L$$

The constraint matrix of this deterministic problem with explicit nonanticipativity shows the structure depicted in Figure 3.2, where again some slack variables' coefficients are neglected.

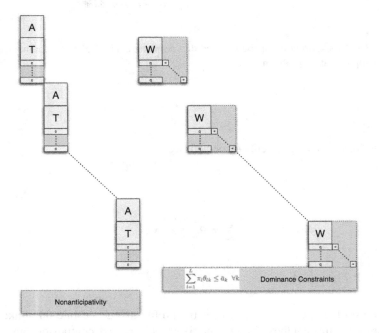

Figure 3.2: Block-structure of the reduced deterministic equivalent for a dominance constrained optimization problem with explicit nonanticipativity

We arrived at an L-shaped form, where all constraints are separable according to the L scenarios except for the two blocks "Nonanticipativity" and "Dominance Constraints'. Relaxing these constraints allows for the separable solution of scenario-specific subproblems instead of the complete deterministic equivalent. A dual decomposition approach which is based on this observation is described below.

The idea behind this approach is an iterative computation of lower bounds and feasible points. The lower bounds are generated by different relaxations of the linking dominance and nonanticipativity constraints. This finally enables the complete decomposition of the original problem into scenario-specific subproblems that can be handled by the solver separately. Lower bounding procedures are established in the following paragraphs. The different scenario-specific first-stage solutions yielding the lower bounds can be understood as a proposal for a feasible solution of the original problem. Subsequently, we present a method to reasonably choose a candidate for a feasible solution and to check its feasibility. The lower and upper bounds are embedded into a branch-and-bound framework, which is described in detail in the last paragraph of this section. The branch-and-bound method is well known in mixed-integer linear programming and serves – in our case – to recover the relaxed dominance and nonanticipativity constraints.

Relaxations and lower bounds

Lagrangian relaxation of the dominance constraints   One possibility for the computation of a lower bound for the original dominance constrained problem is to ignore the explicit nonanticipativity constraints $x_1 = x_2 = \ldots = x_L$ and to apply a *Lagrangian relaxation* of the dominance constraints $\sum_{l=1}^{L} \pi_l \theta_l \leq \bar{a}_k$ for all $k \in 1, \ldots, D$. Considering Figure 3.2, this yields the decomposition of (3.18) into L scenario-specific subproblems.

Obviously, the *Lagrangian function*

$$
\begin{aligned}
\mathscr{L}(x, \theta, \lambda) &= \sum_{l=1}^{L} \pi_l \cdot g^\top x_l + \sum_{k=1}^{D} \lambda_k \Big(\sum_{l=1}^{L} \pi_l \theta_{lk} - \bar{a}_k \Big) \\
&= \sum_{l=1}^{L} \pi_l \cdot g^\top x_l + \sum_{l=1}^{L} \sum_{k=1}^{D} \lambda_k \cdot (\pi_l \theta_{lk} - \pi_l \bar{a}_k) \\
&= \sum_{l=1}^{L} \pi_l \mathscr{L}_l(x_l, \theta_l, \lambda),
\end{aligned}
$$

where

$$\mathscr{L}_l(x_l, \theta_l, \lambda) := g^\top x_l + \sum_{k=1}^{D} \lambda_k \cdot (\theta_{lk} - \bar{a}_k),$$

can be written as the sum of L scenario-specific Lagrangian functions. The arising *Lagrangian dual* then reads

$$\max \left\{ D(\lambda) \, : \, \lambda \in \mathbb{R}_+^D \right\} \tag{3.22}$$

with

$$D(\lambda) = \min \left\{ \begin{array}{rcll} \mathscr{L}(x, \theta, \lambda) \, : & & & \\[1ex] c^\top x_l + q^\top y_{lk} - a_k & \leq & \mathsf{M}\theta_{lk} & \forall l \, \forall k \\[1ex] Tx_l + Wy_{lk} & = & z_l & \forall l \, \forall k \\[1ex] x_l \in X, \ y_{lk} \in \mathbb{Z}_+^{\bar{m}} \times \mathbb{R}_+^{m'}, \ \theta_{lk} \in \{0,1\} & & & \forall l \, \forall k \end{array} \right\}. \tag{3.23}$$

Since $D(\lambda)$ provides a lower bound for (3.18) for each $\lambda \in \mathbb{R}_+$, the Lagrangian dual aims at the computation of the maximum of these lower bounds. $D(\lambda)$ readily decomposes into scenario-specific subproblems, which is illustrated by the following reformulation of (3.23)

$$D(\lambda) = \sum_{l=1}^{L} \pi_l \min \left\{ \begin{array}{rcll} \mathscr{L}_l(x_l, \theta_l, \lambda) \, : & & & \\[1ex] c^\top x_l + q^\top y_{lk} - a_k & \leq & \mathsf{M}\theta_{lk} & \forall k \\[1ex] Tx_l + Wy_{lk} & = & z_l & \forall k \\[1ex] x_l \in X, \ y_{lk} \in \mathbb{Z}_+^{\bar{m}} \times \mathbb{R}_+^{m'}, \ \theta_{lk} \in \{0,1\} & & & \forall k \end{array} \right\}. \tag{3.24}$$

The mixed-integer linear subproblems usually are solvable by standard solvers, whereas the complete optimization problem (3.23), which becomes quite large-scale for big numbers L and D, often is not.

The Lagrangian dual (3.22) is a non-smooth concave maximization (or convex minimization) with piecewise linear objective function which can be solved by iterative bundle methods, for instance see [HK02, Kiw90]. In each step of these methods the objective function value and a subgradient of (3.23) have to be passed to the solver. In case of minimization, a negative (sub-)gradient is given by

$$-\left(\sum_{l=1}^{L} \pi_l \theta_{lk} - \bar{a}_k \right)_{k=1,\dots,D}.$$

As already stressed above, both, the calculation of the objective function value and this subgradient, can exploit the separability of $D(\lambda)$.

Remark 3.24
Certainly, it would be possible to apply Lagrangian relaxation also to the nonanticipativity constraints, which has been done in case of the mean-risk models. Here, the decision to ignore nonanticipativity results from the very time consuming calculation of the Lagrangian dual which directly corresponds to its dimension. Whereas Lagrangian relaxation of the dominance constraints yields a problem of dimension D, the relaxation of nonanticipativity would cause $(m-1) \cdot L$ further Lagrangian multipliers. Moreover, in real world problems the number of scenarios L is often much higher compared to the number of reference profiles D that result from experience or earlier optimization. Hence, to end up with an affordable effort, we forego Lagrangian relaxation of nonanticipativity. Besides, it is much easier to recover the nonanticipativity of the first-stage variables by the branch-and-bound approach than to recover the violation of dominance constraints.

Maximum lower bound of the subproblems If the computational effort for the solution of the Lagrangian dual is very high, it can be reasonable to apply this much simpler lower bound. It is derived from ignoring nonanticipativity as well as the linking dominance constraints. Problem (3.18) then decomposes into very similar scenario-specific subproblems as the above described objective of the Lagrangian dual (3.22). A lower bound is simply given by the maximum optimal solution value of all these mixed-integer linear subproblems.

Let (P_l), $l = 1, \ldots, L$, denote the single-scenario subproblem for scenario l which occurs when both, the nonanticipativity and the linking dominance constraints, are relaxed

$$(P_l) \qquad \min\left\{ g^\top x_l : \begin{array}{rcll} c^\top x_l + q^\top y_{lk} - a_k & \leq & M\theta_{lk} & \forall k \\[2mm] Tx_l + Wy_{lk} & = & z_l & \forall k \\[2mm] x_l \in X, \; y_{lk} \in \mathbb{Z}_+^{\bar{m}} \times \mathbb{R}_+^{m'}, \; \theta_{lk} \in \{0,1\} & & & \forall k \end{array} \right\} \qquad (3.25)$$

and φ_l^{opt} the optimal objective function value of each of these subproblems.

Proposition 3.25
Let (P_l) and φ_l^{opt}, $l = 1, \ldots, L$, be defined as described before. Then

$$\varphi_{LB} := \max\left\{ \varphi_l^{\text{opt}} : l = 1, \ldots, L \right\}$$

is a lower bound for (3.18).

Proof
The proposition is proven by contradiction.
Assume $\varphi_{LB} := \max\left\{\varphi_l^{\mathrm{opt}} : l = 1,\ldots,L\right\}$ is not a lower bound for (3.18). Moreover, let $(P_{\bar{l}})$ be a scenario-specific subproblem with the optimal objective value $\varphi_{\bar{l}}^{\mathrm{opt}} = \varphi_{LB}$. Then there exists a feasible point $(\hat{x},\hat{y},\hat{\theta})$ for (3.18) with

$$g^\top \hat{x} < \varphi_{LB} = \varphi_{\bar{l}}^{\mathrm{opt}}. \qquad (3.26)$$

Since the constraint sets of all (P_l), $l = 1\ldots,L$ are supersets of the feasible set of (3.18), $\hat{x}$ is a feasible first-stage selection for $(P_{\bar{l}})$ as well. This leads to

$$\varphi_{\bar{l}}^{\mathrm{opt}} \le g^\top \hat{x},$$

and we obtain a contradiction to (3.26).   □

The fact that the objective function $g^\top x$ of all subproblems only contains first-stage variables is essential here. In the case of mean-risk models, this lower bound is not applicable because there the objective additionally includes second-stage costs.

Second order dominance constrained problem as a lower bound In Section 3.1.2 we mentioned that first order stochastic dominance implies second order dominance. Hence, if we replace the first order by the second order relation in the stochastic optimization problem with dominance constraints (3.1), we obtain a relaxation of (3.1), namely

$$\min\left\{g(x) : f(x,z) \succeq_{(2)} a, \; x \in X\right\}. \qquad (3.27)$$

As for the first order dominance constrained problem, a deterministic equivalent for (3.27) can be established. We simply present it here and refer to [GGS07] for a proof.

Proposition 3.26
Assume (1.5) – (1.7). Let $z(\omega)$ and $a(\omega)$ in problem (3.27) be discretely distributed with realizations z_l, $l = 1,\ldots,L$, and a_k, $k = 1,\ldots,D$, as well as probabilities π_l, $l = 1,\ldots,L$, and p_k, $k = 1,\ldots,D$, respectively, and let $\mu \in \mathscr{P}(\mathbb{R}^s)$ and $\nu \in \mathscr{P}(\mathbb{R})$ denote the probability measures induced by $z(\omega)$ as well as $a(\omega)$. Moreover, let $g(x) := g^\top x$ be linear and let X

be bounded. Then (3.27) is equivalent to the mixed-integer linear program

$$\min\Big\{g^\top x: \quad c^\top x + q^\top y_{lk} - a_k \;\leq\; s_{lk} \qquad \forall l \,\forall k$$

$$\sum_{l=1}^{L} \pi_l s_{lk} \;\leq\; \hat{a}_k \qquad \forall k$$

$$Tx + W y_{lk} \;=\; z_l \qquad \forall l \,\forall k \Big\}, \qquad (3.28)$$

$$x \in X,\; y_{lk} \in \mathbb{Z}_+^{\bar{m}} \times \mathbb{R}_+^{m'},\; s_{lk} \in \mathbb{R}_+ \quad \forall l \,\forall k$$

where $\hat{a}_k := \int_{\mathbb{R}} \max\{a - a_k, 0\}\, dv,\; k = 1, \ldots, D.$

This mixed-integer linear optimization problem can be solved either directly by standard mixed-integer linear solvers or by one of the lower bounding procedures described above, which again enables the decomposition of (3.28) into scenario-specific subproblems.

Dominance constraints induced by linear recourse Obviously, ignoring integer requirements in the second stage of the optimization problem with first order dominance constraints induced by mixed-integer linear recourse (3.1) yields a further relaxation. The arising optimization problem enables the application of a *cutting plane method*. Let us assume that the value function Φ of the second stage is replaced by

$$\bar{\Phi} := \min\{q^\top y : Wy = t,\; y \in \mathbb{R}_+^{\bar{m}+m'}\}. \qquad (3.29)$$

Moreover, let $(\delta_i, \delta_{i_0}) \in \mathbb{R}^{s+1}$, $i = 1, \ldots, I$, denote the vertices of

$$\{(u, u_0) \in \mathbb{R}^{s+1} : 0 \leq u \leq 1,\; 0 \leq u_0 \leq 1,\; W^\top u - qu_0 \leq 0\}.$$

We obtain that

$$f(x, z) \leq \eta$$

if and only if

$$(z - Tx)^\top \delta_i + (c^\top x - \eta)\delta_{i_0} \leq 0 \quad \forall i.$$

It follows that there exists a constant M such that the problem with first order dominance constraints induced by linear instead of mixed-integer linear recourse (see (3.29)) is equivalent to

$$\min\Big\{g^\top x: \quad (z_l - Tx)^\top \delta_i + (c^\top x - a_k)\delta_{i_0} \;\leq\; M\theta_{lk} \quad \forall l \,\forall k \,\forall i$$

$$\sum_{l=1}^{L} \pi_l \theta_{lk} \;\leq\; \bar{a}_k \qquad \forall k \Big\}, \qquad (3.30)$$

$$x \in X,\; \theta_{lk} \in \{0, 1\} \qquad \forall l \,\forall k$$

where $\bar{a}_k := 1 - v[a \leq a_k]$, $k = 1, \ldots, D$.

This representation gives rise to an iterative cutting plane algorithm, which in each step n solves master problems

$$\min\left\{g^\top x: \begin{array}{rclr} (z_l - Tx)^\top \delta_i + (c^\top x - a_k)\delta_{i_0} & \leq & \mathsf{M}\theta_{lk} & \forall l\ \forall k\ \forall i \in \mathscr{I}_n \\[2ex] \displaystyle\sum_{l=1}^L \pi_l \theta_{lk} & \leq & \bar{a}_k & \forall k \\[2ex] x \in X,\ \theta_{lk} \in \{0,1\} & & & \forall l\ \forall k \end{array}\right\}$$

with $\mathscr{I}_n \subseteq \{i = 1, \ldots, I\}$, and adds violated cuts

$$(z_l - Tx)^\top \delta_i + (c^\top x - a_k)\delta_{i_0} \leq \mathsf{M}\theta_{lk}.$$

The cuts are derived from the dual solutions of linear subproblems. Since the master problems are derived from problem (3.30), which already is a relaxation of (3.1), every iteration of this algorithm provides a lower bound for (3.1). Furthermore, it can be proven that the cutting plane algorithm ends up with an optimal solution of (3.30) after finitely many iterations (see [Dra07, DS07] for details).

Summary on lower bounding procedures Which lower bounding procedure is preferable has to be chosen depending on the concrete optimization problem. Often a combination of methods yields the best results in the shortest computation time. For example, it can be advantageous to use basically the Lagrangian method (in combination with the first or the second order dominance constraints) and additionally the simple lower bound in every k-th node of the branch-and-bound scheme, which is described below.

Also the additional Lagrangian relaxation of nonanticipativity can be reasonable as done for the mean-risk models. On the one hand the dimension of the Lagrangian dual grows and it might become much harder to solve. On the other hand this might provide much better lower bounds. Therefore, the quality of the lower bounds gained from the bounding procedure that ignores nonanticipativity can serve as a criterion to apply Lagrangian relaxation to nonanticipativity and accept a higher computational effort or not.

Finally, the cutting plane method can be a desirable alternative if there are not too many integer requirements in the second stage of the original dominance constrained problem.

In computational tests different lower bounding methods have been compared. The results can be found in Section 4.3.4.

Upper bounds In this paragraph we present a heuristic that aims at finding a feasible point to the original dominance constrained problem (3.18). All lower bounding procedures provide proposals $\tilde{x}_1, \tilde{x}_2, \ldots, \tilde{x}_L$ for a feasible first-stage variable $\tilde{x}$, but a priori they might not fulfill nonanticipativity and the dominance constraints $\sum_{l=1}^{L} \pi_l \theta_{lk} \leq \bar{a}_k$, $k = 1, \ldots, D$. These have to be recovered by the heuristic which is given by the following algorithm.

Algorithm 3.27

STEP 1
Suppose that the solutions $\tilde{x}_1, \tilde{x}_2, \ldots, \tilde{x}_L$ gained from a lower bounding procedure are proposals for the first-stage variable x and generate/choose one "reasonable" (description follows below) $\tilde{x}$ that fulfills all integer requirements in X. Whether $\tilde{x}$ is feasible for problem (3.18), will be checked in the subsequent steps.

STEP 2
Solve the following mixed-integer linear optimization problem for $l = 1, \ldots, L$

$$\min \left\{ \sum_{k=1}^{D} \theta_{lk} : \begin{array}{rcll} c^\top \bar{x} + q^\top y_{lk} - a_k & \leq & M\theta_{lk} & \forall k \\ T\bar{x} + W y_{lk} & = & z_l & \forall k \\ y_{lk} \in \mathbb{Z}_+^{\bar{m}} \times \mathbb{R}_+^{m'}, \ \theta_{lk} \in \{0,1\} & & & \forall k \end{array} \right\}. \tag{3.31}$$

As soon as one of the problems is infeasible, $\tilde{x}$ cannot be feasible for (3.18). The heuristic stops and assigns the formal upper bound $+\infty$. Otherwise, continue with Step 3.

STEP 3
Check for $k = 1, \ldots, D$ whether the $\bar{\theta}_{lk}$ found by the solution of (3.31) fulfill the relaxed dominance constraints

$$\sum_{l=1}^{L} \pi_l \bar{\theta}_{lk} \leq \bar{a}_k.$$

If they do, a feasible solution of (3.18) is found. The heuristic stops with the upper bound $g^\top \bar{x}$.
If not, the heuristic stops without a feasible point for (3.18) and assigns the formal upper bound $+\infty$.

The "reasonable" selection or generation of one first-stage solution $\tilde{x}$ in Step 1 can be done in different ways. On the one hand we can simply choose one of the proposals $\tilde{x}_1, \tilde{x}_2, \ldots, \tilde{x}_L$. This could be the one that occurs most frequently, the one belonging to the scenario with the highest costs, or the one belonging to the scenario with the highest probability, for example. On the other hand we can generate one $\bar{x}$ from the proposals. For instance, this can be the mean value of all proposals. Via this selection/generation the nonanticipativity of the first-stage is recovered. To recover also integer requirements in X, the corresponding coordinates of $\bar{x}$ are rounded to the next integer.

The solution of the mixed-integer programs (3.31) in Step 2 accomplishes two purposes. The first one is to check whether the $\bar{x}$ from Step 1 is feasible for all constraints except the linking dominance constraints of problem (3.18). The second purpose is to find feasible $\bar{\theta}_{lk}$ and to push as many as possible of them to zero. This already aims at recovering the relaxed dominance constraints.

In Step 3 it remains to check whether the $\bar{\theta}_{lk}$ found in Step 2 fulfill the dominance constraints.

A branch-and-bound algorithm   The above described procedures for the computation of lower and upper bounds for problem (3.18) are embedded into a *branch-and-bound scheme*. The idea behind that is to divide the set X with increasing granularity maintaining the linear formulation of the constraint set. Elements of the partition of X, which correspond to nodes of the arising branching tree, can be pruned by optimality, inferiority, or infeasibility. For a more detailed description see Section 2.2.3. This results in Algorithm 3.28 which follows similar lines as the branch-and-bound algorithm for mean-risk models presented in Section 2.2.3.

Let $\mathbf{P}$ denote a list of problems and $\varphi_{LB}(P)$ a lower bound for the optimal value of a problem $P \in \mathbf{P}$. Furthermore, $\bar{\varphi}$ denotes the currently best upper bound to the optimal value of (3.18), and $X(P)$ is the element in the partition of X belonging to P.

Algorithm 3.28

STEP 1 (INITIALIZATION)
Let $\mathbf{P} := \{(3.18)\}$ and $\bar{\varphi} := +\infty$.

STEP 2 (TERMINATION)
If $\mathbf{P} = \emptyset$, then the $\bar{x}$ that yielded $\bar{\varphi} = g^{\top}\bar{x}$ is optimal.

STEP 3 (BOUNDING)
Select and delete a problem P from $\mathbf{P}$. Compute a lower bound $\varphi_{LB}(P)$ using one of the lower bounding procedures described above and find a feasible point $\bar{x}$ for P with Algorithm 3.27.

STEP 4 (PRUNING)
If $\varphi_{LB}(P) = +\infty$ or $\varphi_{LB}(P) > \bar{\varphi}$ (inferiority of P), then go to Step 2.

If $g^\top \bar{x} < \bar{\varphi}$, then $\bar{\varphi} := g^\top \bar{x}$.

If $\varphi_{LB}(P) = g^\top \bar{x}$ (optimality of P), then go to Step 2.

STEP 5 (BRANCHING)
Create two new subproblems by partitioning the set $X(P)$. Add these subproblems to **P** and go to Step 2.

If no feasible solution $\bar{x}$ of (3.18) was found in Step 3, none of the criteria in Step 4 applies and the algorithm proceeds with Step 5.

As described for mean-risk models in Section 2.2.3, the partitioning in Step 5, in principle, can be carried out by adding appropriate linear constraints. This is usually done by branching along coordinates. This means to pick a component $x_{(k)}$ of x and add the inequalities $x_{(k)} \leq a$ and $x_{(k)} \geq a+1$ with some integer a if $x_{(k)}$ is an integer component or, otherwise, add $x_{(k)} \leq a - \varepsilon$ and $x_{(k)} \geq a + \varepsilon$ with some real number a, where $\varepsilon > 0$ is some tolerance parameter to avoid endless branching.

Usually Step 2 is modified using the gap between the objective value induced by the currently best found solution and the currently best lower bound as a stopping criterion for the algorithm. If it falls below some given threshold, the algorithm terminates.

4 Application: Optimal Operation of a Dispersed Generation System

In this chapter we apply the introduced theories and algorithms to an optimization problem from power planning. We consider a *dispersed generation system* which is run by a German utility. We aim at an optimal operation with respect to technical constraints, the supply of thermal an electric demand, and the minimization of operational costs.

4.1 A Dispersed Generation System

A dispersed generation (DG) system is a combination of several power and/or heat generating units with a low capacity compared to conventional nuclear or coal-fired power stations. The single units, which can be installed decentrally, next to the consumers, are linked via communication networks and considered as one power producing system, also called *Virtual Power Plant* ([AKKM02, HBU02, IR02]). A definition can be found in [AAS01]. As reported in former studies (see e.g., [HNNS06]), the operation of dispersed generation units as one system is economically superior to the autarkic operation of each single unit.

Dispersed generation units can be combined heat and power (CHP) units which produce heat and power simultaneously, for example fuel cells, gas motors, or gas turbines, as well as units gaining power from renewable resources like wind turbines, hydroelectric power plants, or photovoltaic devices. Usually, also boilers are included to supply load peaks of heat. Furthermore, DG systems are often equipped with thermal storages and also with cooling devices to exhaust excessive heat. Electric storages are not considered here because either their capacity is too big to be used in DG systems or they store energy only for such a short time that it is not relevant for optimization. Instead, we assume that electric energy can always be sold completely and imported if production does not meet demand.

DG systems gain more and more importance today because of several reasons ([Neu04]). On the one hand they are preferable over custom power plants because of their high overall efficiency – installation next to consumers avoids transportation costs ([BHHU03]) – as well as the comparably low investments needed. They show a high flexibility which enables the operator to react immediately and to supply sudden load peaks, for example. Furthermore, the energy generation with dispersed generation is environmentally friendly compared to convential power generation. On the other hand the current political development promotes the installation of dispersed generation units. For example, the pending nuclear phase-out has to be compensated ("Gesetz zur geordneten Beendigung der Kernener-gienutzung zur gewerblichen Erzeugung von Elektrizität", AtG-E). Moreover, the usage of renewable resources is encouraged by the"Erneuerbare Energien Gesetz" (EEG), where it is stated that power generation from renewable resources should be increased to 20 % of the whole power production in 2020. Furthermore, the "Kraft-Wärme-Kopplungsgesetz" (KWKG) dictates that CO_2 emissions should be reduced by at least 20 million tons until 2010 (based on the emissions of 1998) by an increase of the installed CHP capacity. Additionally, the power generation with CHP units is subsidized. Last but not least many of the existing generation capacities are out-aged and will have to be replaced in the next years ([MMV05]). This stimulates a discussion of new trends, efficiency improvements, and the development of innovative techniques in power generation.

The optimal operation of a DG system requires complex decisions ([Han02]) which are substantially influenced by uncertainty. On the consumers side the heat and power demand as well as the energy prices are not known with certainty and can hardly be predicted. On the production side the infeed from renewables and the fuel costs are stochastic. The data depends on the consumers' behavior, the weather conditions, and the current development of supply and demand on energy markets which cannot be foreseen exactly. Hence, to make reasonable operational decisions all these uncertainties have to be regarded.

The first step to apply stochastic optimization to this management problem is the development of an appropriate mathematical model. In the following section we design such a general mixed-integer linear formulation of the cost minimal operation of a DG-system including all technical and logical constraints.

4.2 Formulation as Linear Mixed-Integer Optimization Problem

We assume that the dispersed generation system consists of W wind turbines, H hydroelectric power plants, and N engine-based cogeneration (CG) stations. Station $n \in \{1, \ldots, N\}$ includes $M[n]$ gas motors, $T[n]$ gas turbines, both producing power and heat simultaneously, and $B[n]$ heat generating boilers. The planning horizon is split into T subintervals, each of length m.

Let U be an arbitrary mentioned production unit. Then p_U^t and h_U^t give the amount of power and heat, respectively, that are produced in time interval $t \in \{1, \ldots, T\}$, and the boolean variable s_U^t indicates whether the unit's production state is on (variable is 1) or off (0) in time interval t. To model minimum up- and down-times for unit U, we need additional boolean variables $u_U^{up\ t}$ and $u_U^{down\ t}$ that are set to 1 if the unit is switched on or to 0 if it is switched off, respectively, in time interval t.

To reflect the usage of the thermal storages, we need a variable sto_n^t which gives the fill of the storage in CG-station n in interval t. Variable in_n^t represents the amount of heat that is injected into this storage in time interval t and out_n^t is the withdrawn amount of heat. Furthermore, the excessive heat which is exhausted by the cooling device in CG-station n in t is given by cl_n^t.

At least there are variables I^t, E_{reg}^t, and E_{fos}^t that denote the power that is imported/bought or exported/sold in time interval t. E_{reg}^t is gained from renewable resources, and E_{fos}^t results from fossil energy sources.

The gas motors, wind turbines, and hydroelectric power plants are working on a maximum level once they have been switched on. Hence, for each gas motor $M_{nj}, n \in \{1, \ldots, N\}, j \in \{1, \ldots, M[n]\}$, each wind turbine $W_j, j \in \{1, \ldots, W\}$, and each hydro power plant $H_j, j \in \{1, \ldots, H\}$, the constraints

$$p_{M_{nj}}^t = s_{M_{nj}}^t \cdot p_{M_{nj}}^{\max},$$

$$h_{M_{nj}}^t = s_{M_{nj}}^t \cdot h_{M_{nj}}^{\max},$$

$$p_{W_j}^t = s_{W_j}^t \cdot p_{W_j}^{\max},$$

$$p_{H_j}^t = s_{H_j}^t \cdot p_{H_j}^{\max}$$

are implemented, where $p^{\max}$ and $h^{\max}$ are the maximum power and heat production levels of the units, respectively.

In contrast, the boilers $B_{nj}, n \in \{1, \ldots, N\}, j \in \{1, \ldots, B[n]\}$, and gas turbines $T_{nj}, n \in \{1, \ldots, N\}, j \in \{1, \ldots, T[n]\}$, can be regulated continuously between a minimum and maximum production level. This is described by

$$s_{B_{nj}}^t \cdot h_{Bnj}^{\min} \leq h_{B_{nj}}^t \leq s_{B_{nj}}^t \cdot h_{Bnj}^{\max},$$

$$s_{T_{nj}}^t \cdot p_{Tnj}^{\min} \leq p_{T_{nj}}^t \leq s_{T_{nj}}^t \cdot p_{Tnj}^{\max},$$

where again $p^{\min}$, $h^{\min}$ and $p^{\max}$, $h^{\max}$ denote minimum and maximum production levels, respectively. The coupled production of heat and power of the gas turbines leads to

$$h_{T_{nj}}^t = \frac{h_{T_{nj}}^{max} - h_{T_{nij}}^{min}}{p_{T_{nj}}^{max} - p_{T_{nj}}^{min}} \cdot \left(p_{T_{nj}}^t - p_{T_{nj}}^{min} \cdot s_{T_{nj}}^t \right) + h_{T_{nj}}^{min} \cdot s_{T_{nj}}^t.$$

To produce energy, the units in the CG-stations consume fuel. This consumption is proportional to the produced power and heat

$$f_{B_{nj}}^t = \left(\frac{f_{B_{nj}}^{max} - f_{B_{nj}}^{min}}{h_{B_{nj}}^{max} - h_{B_{nj}}^{min}} \cdot \left(h_{B_{nj}}^t - h_{B_{nj}}^{min} \cdot s_{B_{nj}}^t \right) + f_{B_{nj}}^{min} \cdot s_{B_{nj}}^t \right) \cdot m,$$

$$f_{M_{nj}}^t = f_{Mnj}^{max} \cdot s_{M_{nj}}^t \cdot m,$$

$$f_{T_{nj}}^t = \left(\frac{f_{T_{nj}}^{max} - f_{T_{nj}}^{min}}{h_{T_{nj}}^{max} - h_{T_{nj}}^{min}} \cdot \left(h_{T_{nj}}^t - h_{T_{nj}}^{min} \cdot s_{T_{nj}}^t \right) + f_{T_{nj}}^{min} \cdot s_{T_{nj}}^t \right) \cdot m.$$

Here, f^{min} and f^{max} are the fuel consumption for minimum and maximum production of each unit, respectively.

In addition to these technical constraints some constraints are needed that logically link the switching variables s_U^t, $u_U^{up\ t}$, and $u_U^{down\ t}$ for each unit U

$$s_U^t - s_U^{t-1} = u_U^{up\ t} - u_U^{down\ t},$$

$$u_U^{up\ t} + u_U^{down\ t} \leq 1.$$

As already mentioned, we guarantee minimum up-times l_U for each unit U with the constraints

$$s_U^{t+i} \geq u_U^{up\ t}, \ \forall t = 1,\ldots,T+1 - l_U, \ i = 0,\ldots,l_U - 1.$$

Furthermore, the storage of heat in each CG-station $n \in \{1,\ldots,N\}$ is modeled by

$$0 \leq in_n^t \leq in_n^{max},$$

$$0 \leq out_n^t \leq out_n^{max},$$

$$sto_n^{min} \leq sto_n^t \leq sto_n^{max},$$

$$sto_n^t = sto_n^{t-1} \cdot \left(1 - \frac{ls}{100} \right) + m \cdot (in_n^t - out_n^t).$$

The amount of heat which is put into or released from the storage is restricted to in_n^{max} and out_n^{max}, respectively, and the fill of the storage has to be in between sto_n^{min} and sto_n^{max}. The last equation links the fill of the storage to the fill in the preceding time interval. A time depending loss ls of heat in the storage is included there.

The amount of excessive heat that must be exhausted through the cooling devices is as well bounded to cl_n^{max}

$$0 \leq cl_n^t \leq cl_n^{max}.$$

Of course, the power that is exported cannot exceed the amount that is produced by the generation units which leads to

$$0 \leq E_{reg}^t \leq \sum_{j=1}^{W} p_{W_j}^t + \sum_{j=1}^{H} p_{H_j}^t,$$

$$0 \leq E_{fos}^t \leq \sum_{i=1}^{N} \left(\sum_{j=1}^{M(n)} p_{M_{nj}}^t + \sum_{j=1}^{T(n)} p_{T_{nj}}^t \right).$$

Finally, there exist constraints that guarantee the supply of the power and heat demand of the consumers. The power demand is given for the whole system for each time interval $t \in \{1, \ldots, T\}$ by D_p^t, whereas the heat is only distributed around every CG-station. Hence, the thermal demand is denoted by a parameter D_h^{tn} for each station $n \in \{1, \ldots, N\}$. The corresponding constraints are

$$D_h^{tn} = \sum_{j=1}^{B(n)} h_{B_{nj}}^t + \sum_{j=1}^{M(n)} h_{M_{nj}}^t + \sum_{j=1}^{T(n)} h_{T_{nj}}^t + out_n^t - in_n^t - cl_n^t,$$

$$D_p^t = \sum_{i=1}^{N} \left(\sum_{j=1}^{M(n)} p_{M_{nj}}^t + \sum_{j=1}^{T(n)} p_{T_{nj}}^t \right) + \sum_{j=1}^{W} p_{W_j}^t + \sum_{j=1}^{H} p_{H_j}^t + I^t - E_{reg}^t - E_{fos}^t.$$

The objective function of the model shall reflect all costs arising from the operation of the dispersed generation system. Therefore, it sums up the fuel costs c^f of gas motors, turbines, and boilers, the start-up costs c^{up} for each unit in the system, and the costs for power import c^I. The incomes from energy export, c_{reg}^E and c_{fos}^E, are included as negative costs

$$
\min \sum_{t=1}^{T} \left[\quad \sum_{n=1}^{N} c^{\text{f}} \cdot \left(\sum_{j=1}^{B(n)} f_{B_{nj}}^{t} + \sum_{j=1}^{M(n)} f_{M_{nj}}^{t} + \sum_{j=1}^{T(n)} f_{T_{nj}}^{t} \right) \right.
$$

$$
+ \quad \sum_{n=1}^{N} c^{\text{up}} \cdot \left(\sum_{j=1}^{B(n)} u_{B_{nj}}^{\text{up}\, t} + \sum_{j=1}^{M(n)} u_{M_{nj}}^{\text{up}\, t} + \sum_{j=1}^{T(n)} u_{T_{nj}}^{\text{up}\, t} \right)
$$

$$
+ \quad \left. \left(c^{\text{I}} \cdot I^{t} - c_{\text{reg}}^{\text{E}} \cdot E_{\text{reg}}^{t} - c_{\text{fos}}^{\text{E}} \cdot E_{\text{fos}}^{t} \right) \cdot m \quad \right].
$$

All constraints and the objective function are linear. Hence, the whole model is a linear mixed-integer representation of the management of the dispersed generation system. The main difficulty rests in boolean variables that represent the operation state and switching operations of each unit.

4.3 Computational Results

In this section computational results for a dispersed generation system run by a German utility are presented. We begin with the deterministic single-scenario problem, proceed with expectation-based and mean-risk extensions, and finally report computations for a model including first order stochastic dominance constraints.

4.3.1 DG configuration and single-scenario problems

The DG system under consideration consists of five engine-based cogeneration stations, twelve wind turbines, and one hydroelectric power plant. The cogeneration stations all in all include eight boilers, nine gas motors, and one gas turbine which sums up to 5.1 MW electric and 39.2 MW thermal capacity. Each CG-station is equipped with a thermal storage and a cooling device, which means that excessive heat can be stored and used in the following time intervals or even be exhausted through the cooling device. Due to the coupled production of heat and power in the gas motors and gas turbine and because of the limited storage, this can become necessary.

The electric energy is fed into the global distribution, whereas the thermal energy is used to supply the heat demand around each CG-station. Due to the substantial loss of thermal energy, the transportation of heat over large distances does not make any sense here. This means we have to meet the heat demand around each CG-station and the global power demand with the operation of the dispersed generation system.

Excessive electric energy cannot be stored, but can be sold like lacking electric energy can be imported. Of course, the energy import is much more expensive than own production. A more detailed description of the system can be found in [Hen06].

In Figure 4.1 the considered DG system is depicted schematically. The first CG-station (BHKW 1), for example, consists of two boilers (KS_11, KS_12), three gas motors (M_11, M_12, M_13), one thermal storage (SP_1), and one cooling device (NK_1), whereas the fifth station (BHKW 5) includes only one boiler (KS_5) and one gas turbine (T_5) instead of gas motors.

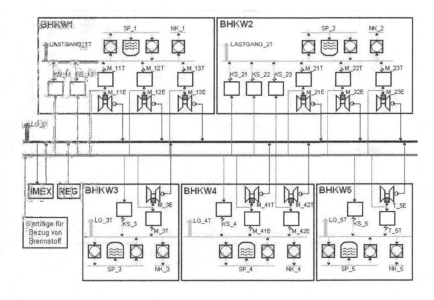

Figure 4.1: Schematic picture of the considered DG system

For the computational tests we consider a planning horizon of 24 hours divided into quarter-hourly subintervals. If we assume the underlying data to be deterministic, which simply means that we choose only one of the scenarios (load profiles, infeed from renewables, and market prices) that could occur, this leads to a deterministic linear mixed-integer optimization problem instance with about 9000 binary variables, 8500 continuous variables, and 22000 constraints.

Such a single-scenario instance is readily solvable with standard software like CPLEX ([ILO05]) which uses LP-based branch-and-bound algorithms. For the computations we use a Linux-PC with a 3.2 GHz Pentium IV processor and 2.0 GB RAM that usually solves the described instances with an optimality gap of 0.1 % in less than 20 seconds.

As an example for a single-scenario solution see Figures 4.2 and 4.3 which display the supply of the power demand for the whole system and the supply of the heat demand around the first CG-station.

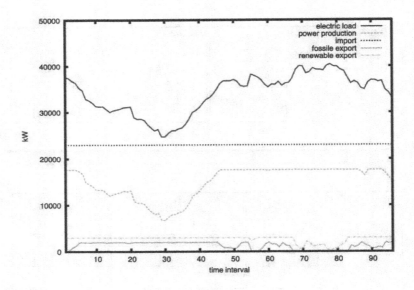

Figure 4.2: Solution for a single-scenario – electric demand, production, import, and export

On the electric side we cannot supply the demand with our own production, which is limited to 17580 kW. Hence, we assume to have an agreement with another utility that allows us to import 23000 kW. In contrast to the description of the model in Section 4.2 some of the gas motors can be operated continuously in between upper and lower bounds. This leads to an exact tracing of the given load profile wherever possible. In time intervals where the imported power exceeds the demand it is sold again. Thereby, the more profitable export from renewable resources is limited to 3000 kW.

According to practical experiences, the operation of a CG-station is determined by the thermal rather than by the electric demand. Therefore, the thermal side of the generation is the more interesting one. For each time interval we have to decide whether the demand is met by production or by heat released from the storage. Usually, it is impossible to meet the demand exactly with production only because of the minimum operation levels of the boilers. The shut down of one boiler leads to a shortage of heat, but even the minimum production of this boiler exceeds demand. Hence, the usage of the storage device is necessary and sensible, though some amount of the stored heat is always lost. All in all, the challenge here is to find a best trade-off between the operation of the boilers and the usage of the storage device.

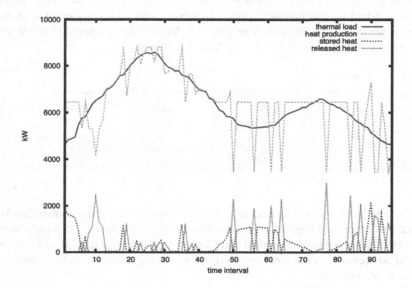

Figure 4.3: Solution for a single-scenario – thermal demand, production, and storage

4.3.2 Expectation-based model

We now include stochasticity into our model. Usually, the input data can be foreseen quite certainly for the time intervals immediately following. In particular, improvements are possible by additional short-term forecasts. However, the prognosis becomes worse the longer the planning horizon is chosen. Therefore, the first four hours of our planning horizon remain deterministic, whereas for the last 20 hours we assume to know the future in form of scenarios with their probabilities to occur. For the development of load profiles see, for example, [GH02]. This information policy fits into the scheme of two-stage stochastic programming with the management decisions of the first four hours as first-stage and the management decisions of the remaining 20 hours as second-stage. As explained in Section 1.1.1, each first-stage decision then defines a random variable $f(x,z)$, see (1.8), which describes the costs induced by x and the realization of $z(\omega)$. Hence, we apply measures to these random variables to find the preferable ones among them. In the simplest case using the expectation of the random costs $\mathbb{E}\big(f(x,z)\big)$ as objective, this yields the expectation-based formulation of the dispersed generation problem.

The deterministic equivalent of formulation (2.7) is again linear mixed-integer, why standard solvers can be applied. However, with a growing number of scenarios the deterministic equivalent gets very large-scale compared to the single-scenario problem in Section 4.3.1, as Table 4.1 shows.

Number of scenarios	Boolean variables	Continuous variables	Constraints
1	8959	8453	22196
5	38719	36613	96084
10	75919	71813	188444
50	373519	353413	927324

Table 4.1: Dimensions of the expectation-based model for 1, 5, 10 and 50 data scenarios

To handle even these large-scale problem instances, we use the above described dual decomposition procedure (see Section 2.2.3). In Table 4.2 a comparison between this decomposition method and CPLEX is reported for test instances with 5 up to 50 scenarios. The stopping criterion for both algorithms is an optimality gap smaller than 0.01 %.

Obviously, the decomposition method outperforms the standard solver when the number of scenarios increases. Whereas CPLEX beats the decomposition for the small instances with five or ten scenarios, the decomposition solves the problem almost 30 times faster than CPLEX, when 50 scenarios are included.

Number of scenarios	CPLEX			Dual Decomposition		
	Objective	Gap (%)	Time (sec.)	Objective	Gap (%)	Time (sec.)
5	7150989.57	< 0.01	10.51	7150903.05	0.0067	7.35
10	5987558.12	< 0.01	110.75	5987518.67	0.0076	131.99
20	5837560.77	< 0.01	3152.66	5837540.90	0.0085	281.27
30	5928533.49	< 0.01	11741.31	5928529.92	0.0090	443.54
40	5833130.04	< 0.01	14980.29	5833181.39	0.0092	649.61
50	5772205.44	< 0.01	23977.86	5772257.78	0.0092	811.64

Table 4.2: Results for the expectation-based model – decomposition compared to CPLEX

4.3.3 Mean-Risk models

The expectation-based model compares the different first-stage decisions and the arising random variables using the mean costs they induce. This does not take into account that two random variables with the same expected costs can be very different from a risk-averse point of view. This is considered if we apply mean-risk models that use a combination of the expectation and a risk measure to evaluate a decision's quality. Subsequently, we consider the Expected Excess, the Excess Probability, and the α-Conditional Value-at-Risk as risk measures.

As the objective function is a weighted sum of the mean costs and the risk measure, mean-risk optimization is a kind of multicriteria optimization. We aim at finding solutions belonging to the efficient frontier (for more details see Section 2.2) which is done via a variation of the weighting factor ρ. From a decision maker's point of view this weighting factor reflects the importance to minimize costs in the mean compared to the importance of avoiding riskiness of a decision and has to be chosen depending on his individual preferences.

At first, results for the Expected Excess over a target η are presented which, as a deviation measure, computes the expected costs exceeding η. Table 4.3 displays computational outcomes for an instance including ten scenarios and four different choices of ρ with a target η of 6807892. If ρ is set to ∞, this means that the pure risk model is considered. The idea is to examine, how the change in ρ affects the value of the expected costs $\mathbb{Q}_{\mathbb{E}}(x)$, the Expected Excess $\mathbb{Q}_{EE_\eta}(x)$, and the scenario specific deterministic costs for each of the ten scenarios. These computations are made on a Linux-PC with a 3.2 GHz Pentium IV processor and 2 GB RAM. The stopping criterion here is an optimality gap smaller than 0.01 %.

Like expected, with growing ρ, which means raising importance of the risk minimization, the expected costs are neglected and increase (from 5987630 to 7185900), whereas the risk measure decreases (from 489812 to 489762). A corre-

Scenario	Weight for risk measure			
number	0.0001	20	10000	∞
1	5828840	5828850	5828800	6807550
2	6808230	6808130	6807950	6807980
3	7787410	7787390	7787320	7787300
4	8766730	8766730	8766710	8766690
5	9746580	9746570	9746550	9746530
6	10726500	10726500	10726400	10726400
7	4849210	4849630	4849290	6697350
8	3870510	3870780	3870900	6447340
9	2893830	2894010	2894700	6389640
10	1948510	1949060	1949900	6074710
$Q_{\mathbb{E}}(x)$	5987630	5987730	5987740	7185900
$Q_{EE_\eta}(x)$	489812	489795	489763	489762

Table 4.3: Results for the mean-risk model with Expected Excess and different risk weights

sponding development can be observed for the single-scenario costs. As the minimization of risk gets more important, the optimization focuses on the scenarios that exceed the cost target η. For example, the costs of scenario 4 and scenario 5 are minimized with growing ρ which reduces the Expected Excess, and the costs of scenario 9 increase because they are minor the target η and do not affect the value of the risk measure.

Further computational tests serve to compare the performance of the described decomposition method and CPLEX for test instances with growing number of scenarios. These tests are made for the Excess Probability over a target η, which describes the probability that costs exceed this given target, as well as for the α-Conditional Value-at-Risk, which reflects the costs of the $(1 - \alpha) \cdot 100\%$ worst cases that can occur in the future.

Both measures allow for decomposition as their deterministic equivalents show a beneficial block-structure without additional coupling between scenarios except the nonanticipativity of the first stage. The dimensions of the arising large-scale equivalents are almost the same as for the expectation-based model reported in Section 4.3.2.

Table 4.4 shows the results for the mean-risk model using Excess Probability. Computations are carried out for instances with 5 up to 50 scenarios. A timelimit of 5 up to 25 minutes is used as stopping criterion.

The results show that it depends on both, the choice of ρ and the number of scenarios, which of the solvers is preferable. For all test instances the model using

Number of scenarios	Time (sec.)	ρ	CPLEX			Dual decomposition		
			$Q_{\mathbb{E}}(x)$	$Q_{EP_\eta}(x)$	Gap (%)	$Q_{\mathbb{E}}(x)$	$Q_{EP_\eta}(x)$	Gap (%)
5	300	0	7150850	1.000	0.0057	7150866	–	0.0062
		0.0001	7150900	0.400	0.0064	7150913	0.400	0.0066
		10000	7150880	0.400	0.0342	7150930	0.400	0.0348
		∞	infeasible		infinity	7447339	0.400	49.990
10	450	0	5987670	1.000	0.0104	5987526	–	0.0078
		0.0001	5987630	0.370	0.0098	5987554	0.370	0.0084
		10000	5987630	0.370	0.0315	5987556	0.370	0.0302
		∞	infeasible		infinity	6128173	0.370	35.130
20	600	0	infeasible		infinity	5837500	–	0.0078
		0.0001	infeasible		infinity	5837560	0.205	0.0089
		10000	infeasible		infinity	5837556	0.205	0.0183
		∞	infeasible		infinity	5978187	0.205	26.830
50	1500	0	infeasible		infinity	5772226	–	0.0087
		0.0001	infeasible		infinity	5772271	0.205	0.0096
		10000	infeasible		infinity	5772287	0.205	0.0151
		∞	infeasible		infinity	5912891	0.265	33.960

Table 4.4: Results for the mean-risk model with Excess Probability

the pure risk measure as objective ($\rho = \infty$) is solved by the decomposition method, admittedly with a gap up to 50 %, whereas CPLEX cannot compute any feasible solution. The same difficulties arise for the mean-risk models including 20 or 50 scenarios, where CPLEX cannot provide any solution in the given time range. Also for the test instance with ten scenarios the decomposition method outperforms CPLEX. There both methods find feasible solutions, but the decomposition reaches a smaller optimality gap. At least, only for the instance with five scenarios CPLEX shows a better performance. It solves the mean-risk model to a smaller optimality gap than the decomposition approach.

The computations for the pure risk model with α-Conditional Value-at-Risk, which are reported in Table 4.5, lead to a similar conclusion. The computations are stopped when the optimality gap drops below 0.01 %. CPLEX reaches this gap faster than the decomposition method for small instances with five or ten scenarios. For all other instances the dual decomposition is the preferred method because it solves these instances up to nine times faster than CPLEX.

4.3.4 First order stochastic dominance constraints

The following part of the results' section deals with first order stochastic dominance constraints applied to the optimal operation of dispersed generation. So far, we judged the feasible first-stage decisions by operational costs they incur. On

Number of	Cplex			Dual decomposition		
scenarios	$Q_{CVaR_\alpha}(x)$	Gap (%)	Time (sec.)	$Q_{CVaR_\alpha}(x)$	Gap (%)	Time (sec.)
5	9.256.610	0,0007	23	9.256.642	0,0010	37
10	9.060.910	0,0033	67	9.060.737	0,0013	79
20	8.522.190	0,0040	504	8.522.057	0,0023	436
30	8.996.950	0,0049	5395	8.996.822	0,0033	594
40	8.795.120	0,0049	7366	8.795.064	0,0039	1038
50	8.557.813	0,0050	9685	8.557.755	0,0039	1286

Table 4.5: Results for the pure risk model with α-Conditional Value-at-Risk

the one hand, using an expectation-based problem formulation, we minimized the expected costs over all possible scenarios. On the other hand with the extension to a mean-risk model, we simultaneously minimized the riskiness of a decision. In both cases, the aim was to find a "best" first-stage solution, and being "best" was defined by the minimum expectation and/or risk measure.

Now the sense of optimization changes a bit. Still we want to control the costs induced by the first-stage decision x, but this decision has not to be the best in the spirit of expected costs or riskiness any longer. Instead, we determine a reference decision and claim that the found solution dominates this given decision. This simply means that we optimize over all feasible operation policies which incur costs that, in a sense of first order stochastic dominance, do not exceed a given benchmark. Hence, the search of a best solution turns into the search of an acceptable, dominating solution.

For a realistic test the benchmark profile is a discretely distributed random variable with a limited number of given outcomes and their probability. To derive such a profile, we use an optimal solution $\hat{x}$ found for the expectation-based model from Section 4.3.2 and the corresponding scenario costs $f(\hat{x}, z)$. The benchmark values are heuristically selected and the $f(\hat{x}, z)$ are clustered around them. The probability of each benchmark value then arises as the sum of the probabilities of the members in its cluster. Further problem instances are designed by fixing the probabilities and increasing the values of the benchmark profile successively.

Minimizing abrasion of DG-units   As controlling costs is now part of the constraints and no longer included in the objective, we can additionally formulate other requirements to the found solution by a new, customized objective function. The only restriction is that the objective has to remain linear in the first-stage variables. Obviously, there are many different aspects that could be included into the model this way. One could trace a given schedule in the first-stage, minimize the energy import to be independent from competitors, or maximize the usage of a technology that is eminently environmentally friendly.

The idea we want to discuss here is to minimize the number of start-up processes of generation units in the first-stage time intervals. Then the new objective function looks as follows

$$g(x) := \sum_{t=1}^{16} \left[\sum_{n=1}^{N} \left(\sum_{j=1}^{B(n)} u_{B_{nj}}^{up\ t} + \sum_{j=1}^{M(n)} u_{M_{nj}}^{up\ t} + \sum_{j=1}^{T(n)} u_{T_{nj}}^{up\ t} \right) + \sum_{j=1}^{W} u_{W_j}^{up\ t} + \sum_{j=1}^{H} u_{H_j}^{up\ t} \right].$$

This corresponds to a minimization of the abrasion of generation units because their durability, especially for the CHP units, is substantially influenced by the number of start-ups (for details see [Lei05, SK96]).

We report results for dominance constrained problems, including this objective, with $K = 4$ benchmark profiles and $L = 10$ up to 50 scenarios for heat and power demand. The dimensions of deterministic equivalents according to Section 3.2.3 are displayed in Table 4.6.

Number of scenarios	Boolean variables	Continuous variables	Constraints
10	299159	283013	742648
20	596799	564613	1481568
30	894439	846213	2220488
50	1489719	1409413	3698328

Table 4.6: Dimensions of the first order dominance constrained model

In Tables 4.7 - 4.10 we compare results obtained from the standard mixed-integer solver CPLEX ([ILO05]) to results computed with an implementation of the decomposition Algorithm 3.28 derived in Section 3.2.4. Lower bounds are computed using Lagrangian relaxation of the interlinking first order dominance constraints and ignoring nonanticipativity. Again, computations are done on a Linux-PC with a 3.2 GHz Pentium IV processor and 2 GB RAM and a timelimit of eight hours is used.

We apply five different benchmark profiles for each set of scenarios. These profiles are characterized in the first three columns of the tables together with the probabilities of their benchmark values, numbered instance 1 to instance 5. From 1 to 5 the benchmark costs increase successively which means that the dominance constraints get easier to fulfill. As one would expect, this affects the needed number of start-ups positively. They decrease with increasing reference values (for example, for the ten scenario tests the needed start-ups are 29 - 28 - 21 - 13 - 8), which is reported in the column *Upper Bound*, where the objective function value of the currently best solution is displayed. The corresponding best lower bound is reported in column *Lower Bound*.

Inst.	Benchmarks		Time (sec.)	CPLEX		Dual Decomposition	
	Prob.	Benchmark Value		Upper Bound	Lower Bound	Upper Bound	Lower Bound
1	0.12	2895000	430.43	–	29	29	15
	0.21	4851000	899.16	–	29	29	29
	0.52	7789000	15325.75	29	29	29	29
	0.15	10728000					
2	0.12	2900000	192.48	–	27	28	15
	0.21	4860000	418.90	28	28	28	15
	0.52	7800000	802.94	28	28	28	28
	0.15	10740000					
3	0.12	3000000	144.63	–	21	21	12
	0.21	5000000	428.61	21	21	21	18
	0.52	8000000	678.79	21	21	21	21
	0.15	11000000					
4	0.12	3500000	164.34	–	11	13	10
	0.21	5500000	818.26	–	12	13	13
	0.52	8500000	28800.00	13	12	13	13
	0.15	11500000					
5	0.12	4000000	171.52	–	7	8	8
	0.21	6000000	3304.02	8	8	8	8
	0.52	9000000					
	0.15	12000000					

Table 4.7: Results for instances of a dominance constrained model with ten data and four benchmark scenarios

For all test instances the dual decomposition algorithm reaches the first feasible solution faster than CPLEX does. The time the decomposition needs to find this first feasible solution is thus reported. Furthermore, if CPLEX or the decomposition method is able to solve test instances to optimality, the points in time are given where optimality is proven (optimality gap is zero). For CPLEX this usually turned out to be the time when the first feasible solution was found. Finally, if optimality cannot be proven within eight hours by one of the solvers, the current status of the computation at expiry of the timelimit is reported.

Moreover, some special cases occur. For two of the test instances with 20 scenarios and all test instances including 30 scenarios CPLEX aborts computation before a first feasible solution is reached because the available memory is exceeded

Inst.	Benchmarks		Time (sec.)	CPLEX		Dual Decomposition	
	Prob.	Benchmark Value		Upper Bound	Lower Bound	Upper Bound	Lower Bound
1	0.105	2895000	306.89	–	29	29	12
	0.1	4851000	1151.95	–	29	29	29
	0.69	7789000	9484.97	29	29	29	29
	0.105	10728000					
2	0.105	2900000	703.91	–	27	28	18
	0.1	4860000	1744.75	28	28	28	26
	0.69	7800000	1916.06	28	28	28	28
	0.105	10740000					
3	0.105	3000000	305.84	–	21	21	10
	0.1	5000000	1682.93	21	21	21	19
	0.69	8000000	2138.94	21	21	21	21
	0.105	11000000					
4	0.105	3500000	425.98	–	11	13	9
	0.1	5500000	2213.08	–	12	13	13
	0.69	8500000	11236.31	–	12 mem.	13	13
	0.105	11500000					
5	0.105	4000000	447.33	–	8	8	8
	0.1	6000000	5599.99	9	8	8	8
	0.69	9000000	7840.09	9	8 mem.	8	8
	0.105	12000000					

Table 4.8: Results for instances of a dominance constrained model with 20 data and four benchmark scenarios

(marked by *mem.*). Only the lower bounds already found before the memory error occurred are displayed then. Using 50 data scenarios, the deterministic equivalents get so large that the available memory is even not sufficient to build up the model (lp-) file needed as input for CPLEX which prevents optimization with CPLEX for these instances. Hence, in the last table only best values and lower bounds calculated with the dual decomposition method are presented.

All in all, the computations show that for most purposes the decomposition method is preferable over the standard mixed-integer linear programming solver CPLEX. In all tests it provides a first feasible solution faster then CPLEX (up to 35 times faster), and in most cases this is already an optimal solution of the test problem. Furthermore, usually less computing time is needed to prove optimality of a found solution by improvement of the lower bounds. Only for the test instances 2 and 3, using 10 or 20 scenarios, CPLEX provides tighter lower bounds than the decomposition method.

| Inst. | Benchmarks | | Time (sec.) | CPLEX | | Dual Decomposition | |
	Prob.	Benchmark Value		Upper Bound	Lower Bound	Upper Bound	Lower Bound
1	0.085	2895000	473.27	–	28	29	12
	0.14	4851000	1658.02	–	29	29	29
	0.635	7789000	3255.99	–	29 mem.	29	29
	0.14	10728000					
2	0.085	2900000	1001.53	–	26	28	18
	0.14	4860000	2694.93	–	27	28	28
	0.635	7800000	3372.24	–	27 mem.	28	28
	0.14	10740000					
3	0.085	3000000	469.93	–	17	23	10
	0.14	5000000	3681.15	–	18 mem.	21	20
	0.635	8000000	28800.00	–	–	21	20
	0.14	11000000					
4	0.085	3500000	618.21	–	10	14	8
	0.14	5500000	3095.02	–	11 mem.	14	10
	0.635	8500000	28800.00	–	–	14	13
	0.14	11500000					
5	0.085	4000000	672.73	–	7	8	8
	0.14	6000000	8504.88	–	8 mem.	8	8
	0.635	9000000					
	0.14	12000000					

Table 4.9: Results for instances of a dominance constrained model with 30 data and four benchmark scenarios

The superiority of the dual decomposition over general-purpose solvers gets evident in the computations dealing with 30 and 50 scenarios. For instances with 30 scenarios CPLEX cannot provide any feasible solution, for 50 scenarios even no lower bound, whereas the decomposition algorithm solves these problems. It always provides good solutions, whose optimality can be proven for three out of five instances.

Comparison of lower bounding procedures　As described in Section 3.2.4, there are several possibilities for lower bounding procedures. One of them uses the solution of the stochastic optimization problem with second order dominance constraints (3.28). This lower bounding method shall be compared to the "standard" procedure applying Lagrangian relaxation to the first order dominance constraints, which has been used throughout the above computations. Both lower bounding methods allow for a decomposition, which is enabled by Lagrangian relaxation of

Inst.	Benchmarks		Time (sec.)	CPLEX		Dual Decomposition	
	Prob.	Benchmark Value		Upper Bound	Lower Bound	Upper Bound	Lower Bound
1	0.09	2895000	745.87	–	–	29	11
	0.135	4851000	2534.21	–	–	29	29
	0.67	7789000					
	0.105	10728000					
2	0.09	2900000	1549.22	–	–	28	18
	0.135	4860000	4168.89	–	–	28	28
	0.67	7800000					
	0.105	10740000					
3	0.09	3000000	756.06	–	–	23	10
	0.135	5000000	28800.00	–	–	21	20
	0.67	8000000					
	0.105	11000000					
4	0.09	3500000	975.20	–	–	15	8
	0.135	5500000	28800.00	–	–	13	12
	0.67	8500000					
	0.105	11500000					
5	0.09	4000000	1150.95	–	–	8	8
	0.135	6000000					
	0.67	9000000					
	0.105	12000000					

Table 4.10: Results for instances of a dominance constrained model with 50 data and four benchmark scenarios

the interlinking dominance constraints and the ignorance of nonanticipativity.

Computations have been made for the above described problem considering an optimal operation of a dispersed generation system with minimum number of units' start-ups and overall operation costs that dominate a given benchmark. It turns out that none of the lower bounding methods is decidedly preferable to the other. Obviously, the superiority of a method strongly depends on the concrete problem data. For an illustration see Tables 4.11 and 4.12.

They show the development of both lower bounds over time for problem instance 2 and 4, respectively. The number of included scenarios increases from 20 to 50 for each instance and the four benchmark profiles are supposed the same as above. By (*opt*) we mark the point, where a lower bound has reached the optimal value of the corresponding first or second order problem, respectively, and "-" indicates that no lower bound has been computed yet.

For instance 2 the dual decomposition applied to the first order dominance con-

Number of scenarios	Time (sec.)	Lower bounds provided by	
		First Order Model	Second Order Model
20	377	12	–
	464	12	9
	704	18	9
	977	18	17
	1027	25	17
	1435	25	25
	1444	26	25
	1765	27	25
	1915	28 (opt)	25
	2500	28	27 (opt)
30	529	11	–
	702	11	9
	1002	18	9
	1433	18	17
	1450	25	17
	2100	25	25
	2497	27	25
	2692	28 (opt)	25
	3633	28	27 (opt)
50	813	11	–
	1125	11	9
	1549	18	9
	2208	18	17
	2235	25	17
	3198	25	25
	3853	27	25
	4165	28 (opt)	25
	5854	28	27 (opt)

Table 4.11: Different lower bounds over time for instance 2

strained problem provides the tighter bounds, whereas for instance 4 the lower bounding procedure using the second order dominance constrained problem should be preferred.

Number of scenarios	Time (sec.)	Lower bounds provided by	
		First Order Model	Second Order Model
20	386	–	9
	425	9	9
	789	9	10
	1195	9	11 (opt)
	2211	13 (opt)	11
30	500	–	9
	618	8	9
	1016	8	10
	1402	8	11 (opt)
	3107	10	11
	3566	11	11
	4557	12	11
	5645	13	11
50	975	8	–
	1026	8	9
	2075	8	10
	2333	9	10
	2869	9	11 (opt)
	3456	10	11
	4991	11	11
	6475	12	11

Table 4.12: Different lower bounds over time for instance 4

5 Conclusion and Perspective

In this thesis we considered two fundamental approaches to handle mixed-integer linear optimization problems with uncertainty and applied them to the complex problem of operating a dispersed generation system with minimum costs supplying the demanded heat and power. Both approaches are suitable to reflect the preferences of a rational decision maker, who wants to minimize losses and simultaneously avoid risks.

The first approach uses mean-risk models including different risk measures – Expected Excess, Excess Probability, and α-Conditional Value-at-Risk. We showed well-posedness of these models, reported stability results concerning perturbations of the underlying probability measures, and gave equivalent, deterministic problem formulations. A dual decomposition method to gain lower bounds was described, and an algorithm based on branch-and-bound using these lower bounds and feasibility heuristics was presented.

In a second part of this work we dealt with a new class of optimization problems which also handle two-stage stochastic optimization. Dominance relations were defined and used to establish stochastic optimization problems with first order dominance constraints induced by mixed-integer recourse. We pointed out the special structural properties of these problems, proved their well-posedness, and analyzed their stability if the underlying probability measures are described by approximations. A qualitative stability result could be proven which allows for the usage of approximations to solve the dominance constrained problems. We gave a derivation of the arising deterministic equivalents and developed a tailored solution algorithm, again based on a branch-and-bound scheme. A heuristic to provide feasible solutions was presented and different lower bounding procedures exploiting the special structure of the constraint matrix were established.

The theoretical part of the thesis was followed by the application of both optimization approaches to the optimal operation of a dispersed generation system run by a German utility. Therefore, an existing implementation of the dual decomposition for mean-risk models was used and extended to a solver for dominance constrained problems using special lower bounds and heuristics for this problem class. Furthermore, we developed a mixed-integer model to reflect the operation of dispersed generation with all its technical and economical constraints. Computational results for single-scenario instances and the traditional expectation-based optimization problem pointed out how energy optimization benefits from stochastic programming. On the one hand for complex situations operational schedules with an optimal trade-off between production and storage usage for each single future scenario are provided. On the other hand all predicted scenarios can be included, and solutions which are "best" for all these possible future developments can be found that could not simply be concluded from the single-scenario solutions. Furthermore, we could prove the superiority of tailored decomposition methods over standard solvers for both concepts, the mean-risk approach and the dominance constrained optimization problems. Especially, if the number of included scenarios increases, the usage of decomposition is necessary since standard solvers are carried to their limits. Many problem instances could be solved to optimality with decomposition methods, whereas the standard solver CPLEX could not provide any feasible solution.

The mean-risk models as well as the first order dominance constrained problems turned out to be very flexible and adaptable. The inclusion of different risk measures into mean-risk optimization enables the reflection of each decision maker's personal perception of risk. The application of a dominance relation allows for the definition of an "acceptable" solution in the sense that a solution has to dominate a given cost benchmark, which can be chosen according to a decision maker's desires. Moreover, a new objective function not considering the operational costs can be included into the dominance constrained model since the controlling of costs is moved into the problem's constraints. This provides two possibilities to adjust the dominance constrained optimization problem to the personal preferences of a decision maker.

We already gave an extensive analysis of first order dominance constrained optimization problems here. It remains to prove a quantitative stability result which allows for a statement concerning the deviation of an approximative solution from the exact optimum of these problems. Such a quantitative result should at least be provable for problems without integer requirements in the second stage. The proof should be future work based on results presented in this thesis. Moreover, an enhanced analysis of second order dominance constrained problems would be reasonable and could provide additional insights usable for innovations in energy optimization.

Appendix

Appendix

List of Tables

List of Figures

Bibliography

[AAS01]   Th. Ackermann, G. Andersson, and L. Söder. Distributed Generation: A defi-
            nition. *Electric Power System Research*, 57, 2001.

[AKKM02]   U. Arndt, D. Köhler, Th. Krammer, and H. Mühlbacher. *Das Virtuelle
            Brennstoffzellen-Kraftwerk.* Forschungsstelle für Energiewirtschaft München,
            2002.

[All53]   M. Allais. Le comportement de l'homme rationnel devant le risque: Critique
            des postulats et axiomes de l'ecole americaine. *Econometrica*, 21:503–546,
            1953.

[Baw82]   V. S. Bawa. Stochastic dominance: A research bibliography. *Management
            Science*, 28:698–712, 1982.

[Bea55]   E.M.L. Beale. On minimizing a convex function subject to linear inequalities.
            Journal of the Royal Statistical Society, Series B, 17:173–184, 1955.

[Bea61]   E.M.L. Beale. The use of quadratic programming in stochastic linear program-
            ming. *Rand Report P-2404, The RAND Corporation*, 1961.

[Ben01]   H. P. Benson. Multi-objective optimization: Pareto optimal solutions, proper-
            ties. In C. A. Floudas and P. M. Pardalos, editors, *Encyclopedia of Optimiza-
            tion*, volume III, pages 489–493. Kluwer Academic Publishers, Dordrecht,
            The Netherlands, 2001.

[Ber63]   C. Berge. *Topological Spaces.* Macmillan, New York, 1963.

[Ber75]   B. Bereanu. Stable stochastic integer programs and applications. *Mathema-
            tische Operationsforschung und Statistik*, 6:593–607, 1975.

[BGK+82]   B. Bank, J. Guddat, D. Klatte, B. Kummer, and K. Tammer. *Non-linear Para-
            metric Optimization.* Akademie-Verlag, Berlin, 1982.

[BHHU03]   R. Becker, E. Handschin, E. Hauptmeier, and F. Uphaus. Heat-controlled com-
            bined cycle units in distribution networks. *CIRED 2003*, Barcelona, 2003.

[Bil68]   P. Billingsley. *Convergence of Probability Measures.* Wiley, New York, 1968.

[Bil86]   P. Billingsley. *Probability and Measure.* Wiley, New York, 1986.

[BJ77]   C. E. Blair and R. G. Jeroslow. The value function of a mixed integer program:
            I. *Discrete Mathematics*, 19:121–138, 1977.

[BL97]   J. R. Birge and F. Louveaux. *Introduction to Stochastic Programming.*
            Springer, New York, 1997.

[Bla51]   D. Blackwell. Comparison of experiments. In *Proceedings of the Second Berkeley Symposium on Mathematical Statistics and Probability*, pages 93–102. University of California Press, Berkeley, 1951.

[Bla53]   D. Blackwell. Equivalent comparisons of experiments. *The annals of mathematical statistics*, 24:264–272, 1953.

[BM88]   B. Bank and R. Mandel. *Parametric Integer Optimization*. Akademie-Verlag, Berlin, 1988.

[BW86]   J. R. Birge and J.-B. Wets. Designing approximation schemes for stochastic optimization problems, in particular for stochastic programs with recourse. *Mathematical Programming Study*, 27:54–102, 1986.

[CC59]   A. Charnes and W. W. Cooper. Chance-constrained programming. *Management Science*, 6:73–79, 1959.

[CS99]   C. C. Carøe and R. Schultz. Dual decomposition in stochastic integer programming. *Operations Research Letters*, 24:37–45, 1999.

[Dan55]   G. B. Dantzig. Linear programming under uncertainty. *Management Science*, 1(3/4), 1955.

[Dan67]   J. M. Danskin. *The Theory of Max-Min and its Application to Weapons Allocations Problems*. Springer, New York, 1967.

[DFS67]   G. B. Dantzig, J. Folkman, and N. Shapiro. On the continuity of the minimum set of a continuous function. *Journal of Mathematical Analysis and Applications*, 17:519–548, 1967.

[DHR07]   D. Dentcheva, R. Henrion, and A. Ruszczyński. Stability and sensitivity of optimization problems with first order stochastic dominance constraints. *SIAM Journal on Optimization*, 18(1):322–337, 2007.

[DM61]   G. B. Dantzig and A. Madansky. On the solution of two-stage linear programs under uncertainty. In Y. J. Neyman, editor, *Proc. 4th Berkeley Symp. Math. Stat. Prob., Berkeley*, pages 165–176, 1961.

[DR03a]   D. Dentcheva and A. Ruszczyński. Optimization under Linear Stochastic Dominance. *Comptes Rendus de l'Academie Bulgare des Sciences*, 56:5–10, 2003.

[DR03b]   D. Dentcheva and A. Ruszczyński. Optimization under Nonlinear Stochastic Dominance Constraints. *Comptes Rendus de l'Academie Bulgare des Sciences*, 56:17–22, 2003.

[DR03c]   D. Dentcheva and A. Ruszczyński. Optimization with stochastic dominance constraints. *SIAM Journal on Optimization*, 14(2):548–566, 2003.

[DR04a]   D. Dentcheva and A. Ruszczyński. Optimality and duality theory for stochastic optimization problems with nonlinear dominance constraints. *Mathematical Programming*, 99(2):329–350, 2004.

[DR04b]   D. Dentcheva and A. Ruszczyński. Semi-infinite probabilistic optimization:
           first order stochastic dominance constraints. *Optimization*, 53:583–601, 2004.

[DR06]   D. Dentcheva and A. Ruszczyński. Portfolio optimization with stochastic
           dominance constraints. *Journal of Banking and Finance*, 30:433–451, 2006.

[Dra07]   D. Drapkin. *A Decomposition Algorithm for Two-Stage Stochastic Programs
           with Dominance Constraints Induced by Linear Recourse*. Diploma Thesis,
           University of Duisburg-Essen, Department of Mathematics, 2007.

[DS07]   D. Drapkin and R. Schultz. An algorithm for stochastic programs with first-
           order dominance constraints induced by linear recourse. *Preprint Series, De-
           partment of Mathematics, University of Duisburg-Essen*, 653-2007, 2007.

[Dud89]   R. M. Dudley. *Real Analysis and Probability*. Wadsworth and Brooks/Cole,
           Pacific Grove, CA, 1989.

[Ehr00]   M. Ehrgott. *Multicriteria Optimization*. Springer-Verlag, Berlin, 2000.

[Fis64]   P. C. Fishburn. *Decision and Value Theory*. Wiley, New York, 1964.

[Fis70]   P. C. Fishburn. *Utility Theory for Decision Making*. Wiley, New York, 1970.

[GGNS07]   R. Gollmer, U. Gotzes, F. Neise, and R. Schultz. Risk modeling via stochas-
           tic dominance in power systems with dispersed generation. *Preprint Series,
           Department of Mathematics, University of Duisburg-Essen*, 651-2007, 2007.

[GGS07]   R. Gollmer, U. Gotzes, and R. Schultz. Second-order stochastic dominance
           constraints induced by mixed-integer linear recourse. *Preprint Series, Depart-
           ment of Mathematics, University of Duisburg-Essen*, 644-2007, 2007.

[GH02]   B. Geiger and M. Hellwig. *Entwicklung von Lastprofilen für die Gaswirtschaft
           - Haushalte*. im Auftrag des Bundesverbandes der deutschen Gas- und Wasser-
           wirtschaft, www.bundesverband-gas-und-wasser.de, München, 2002.

[GNS06]   R. Gollmer, F. Neise, and R. Schultz. Stochastic programs with first-order
           dominance constraints induced by mixed-integer linear recourse. *Preprint Se-
           ries, Department of Mathematics, University of Duisburg-Essen*, 641-2006,
           2006. accepted for publication in SIAM Journal on Optimization.

[Han02]   E. Handschin. Das virtuelle Kraftwerk für die Zukunft. *Proceedings of the
           VDI-GET Conference Entwicklungslinien der Energietechnik*, Bochum, 2002.

[HBU02]   E. Handschin, R. Becker, and F. Uphaus. Internet control for decentralized
           energy conversion systems. *2nd International Symposium on Distributed Gen-
           eration*, Stockholm, 2002.

[Hen06]   E. Hennig. Betriebserfahrungen mit dem virtuellen Kraftwerk Unna. *BWK*,
           7/8:28–30, 2006.

[HK02]   C. Helmberg and K. C. Kiwiel. A spectral bundle method with bounds. *Math-
           ematical Programming*, 93:173–194, 2002.

[HL69]   G. Hanoch and H. Levy. The efficiency analysis of choices involving risk.
            Review of Economic Studies, 36:335–346, 1969.

[HLP59]   G. H. Hardy, J. E. Littlewood, and G. Polya. *Inequalities*. University Press
            Cambridge, 1959.

[HNNS06]   E. Handschin, F. Neise, H. Neumann, and R. Schultz. Optimal operation
            of dispersed generation under uncertainty using mathematical programming.
            International Journal of Electrical Power and Energy Systems, 28:618–626,
            2006.

[HR69]   J. Hadar and W. R. Russell. Rules for ordering uncertain prospects. *The
            American Economic Review*, 59:25–34, 1969.

[HR71]   J. Hadar and W. R. Russell. Stochastic dominance and diversification. *Journal
            of Economic Theory*, 3:288–305, 1971.

[HR99]   R. Henrion and W. Römisch. Metric regularity and quantitative stability in
            stochastic programs with probabilistic constraints. *Mathematical Program-
            ming*, 84:55–88, 1999.

[ILO05]   ILOG. CPLEX Callable Library 9.1.3, 2005.

[IR02]   T. Ichikawa and Ch. Rehtanz. Recent trends in distributed generation - tech-
            nology, grid integration, system operation. *Proceedings of 14th Power Systems
            Computation Conference*, Sevilla, 2002.

[Kal74]   P. Kall. Approximations to stochastic programs with complete recourse. *Nu-
            merische Mathematik*, 22:333–339, 1974.

[Kal87]   P. Kall. On approximations and stability in stochastic programming. In J. Gud-
            dat, H. Th. Jongen, B. Kummer, and F. Nožička, editors, *Parametric Optimiza-
            tion and Related Topics*. Akademie-Verlag, Berlin, 1987.

[Kan77]   V. Kanková. Optimum solution of a stochastic optimization problem with
            unknown parameters. *Transactions of the 7th Prague Conference on Informa-
            tion Theory, Statistical Decision Functions and Random Processes, Academia*,
            Prague:239–244, 1977.

[Kan78]   V. Kanková. Stability in the stochastic programming. *Kybernetika*, 14:339–
            349, 1978.

[Kar32]   J. Karamata. Sur une inegalité relative aux fonctions convexes. *Publ. Math.
            Univ. Belgrade*, 1:145–148, 1932.

[Kiw90]   K. C. Kiwiel. Proximity control in bundle methods for convex nondifferen-
            tiable optimization. *Mathematical Programming*, 46:105–122, 1990.

[Kla87]   D. Klatte. *A note on quantitative stability results in nonlinear optimization*.
            Seminarbericht 90, Sektion Mathematik, Humboldt-Universität zu Berlin,
            Berlin, 1987.

[Kla94]   D. Klatte. On quantitative stability for non-isolated minima. *Control and Cybernetics*, 23:183–200, 1994.

[KM05]   P. Kall and J. Mayer. *Stochastic Linear Programming*. International Series in Operations Research and Management Science. Springer, New York, 2005.

[Kor01]   P. Korhonen. Multi objective programming support. In C. A. Floudas and P. M. Pardalos, editors, *Encyclopedia of Optimization*, volume III, pages 566–574. Kluwer Academic Publishers, Dordrecht, The Netherlands, 2001.

[KW94]   P. Kall and S. W. Wallace. *Stochastic Programming*. Wiley, Chichester, 1994.

[Leh55]   E. Lehmann. Ordered families of distributions. *The annals of mathematical statistics*, 26:399–419, 1955.

[Lei05]   P. Leijendeckers. Kraft-Wärme-Kopplung nach VDI 3985 - Einführung. *VDI Berichte*, 1881, 2005.

[Lev92]   H. Levy. Stochastic dominance and expected utility: Survey and analysis. *Management Science*, 38(4):555–593, 1992.

[Mar52]   H. M. Markowitz. Portfolio selection. *Journal of Finance*, 7:77–91, 1952.

[Mar59]   H. M. Markowitz. *Portfolio Selection*. John Wiley and Sons, New York, 1959.

[Mar75]   K. Marti. Approximationen der Entscheidungsprobleme mit linearer Ergebnisfunktion und positiv homogener, subadditiver Verlustfunktion. *Zeitschrift für Wahrscheinlichkeitstheorie und verwandte Gebiete*, 31:203–233, 1975.

[Mar79]   K. Marti. Approximationen stochastischer Optimierungsprobleme. In *Mathematical Systems in Economics*, volume 43. A. Hain, Königstein/Ts, 1979.

[Mar87]   H. M. Markowitz. *Mean-Variance Analysis in Portfolio Choice and Capital Markets*. Blackwell, Oxford, 1987.

[MMV05]   P. Markewitz, D. Martinsen, and S. Vögele. Mögliche Entwicklungen und Auswirkungen eines zukünftigen Kraftwerksbedarfs. *Zeitschrift für Energiewirtschaft*, 2005-3, 2005.

[MS91]   K. C. Mosler and M. Scarsini, editors. *Stochastic orders and decision under risk*. Institute of Mathematical Statistics, Hayward, CA, 1991.

[MS02]   A. Müller and D. Stoyan. *Comparison Methods for Stochastic Models and Risks*. John Wiley and Sons, Chichester, UK, 2002.

[Neu04]   H. Neumann. Zukünftige Infrastruktur der verteilten Energieversorgung. *Proceedings of the Conference Fern-/Nahwärme und Kraft-Wärme-Kopplung-Energie für Menschen mit Weitblick*, Leipzig, 2004.

[Neu07]   H. Neumann. *Zweistufige stochastische Betriebsoptimierung eines Virtuellen Kraftwerks*. Shaker Verlag, Aachen, July 2007.

[NR08]   N. Noyan and A. Ruszczyński. Valid inequalities and restrictions for stochastic programming problems with first order stochastic dominance constraints. *Mathematical Programming*, 114(2):249–275, 2008.

[NRR06]   N. Noyan, G. Rudolf, and A. Ruszczyński. Relaxations of linear program-
          ming problems with first order stochastic dominance constraints. *Operations
          Research Letters*, 34:653–659, 2006.

[NW88]   G. L. Nemhauser and L. A. Wolsey. *Integer and Combinatorial Optimization*.
          Wiley, New York, 1988.

[Ols76]   P. Olsen. Discretizations of multistage stochastic programming problems.
          Mathematical Programming Study, 6:111–124, 1976.

[OR99]   W. Ogryczak and A. Ruszczyński. From stochastic dominance to mean-risk
          models: Semideviations as risk measures. *European Journal of Operations
          Research*, 116:33–50, 1999.

[OR01]   W. Ogryczak and A. Ruszczyński. On consistency of stochastic dominance
          and mean-semideviation models. *Mathematical Programming*, 89:217–232,
          2001.

[OR02]   W. Ogryczak and A. Ruszczyński. Dual stochastic dominance and related
          mean-risk models. *SIAM Journal on Optimization*, 13:60–78, 2002.

[Pol84]   D. Pollard. *Convergence of Stochastic Processes*. Springer-Verlag, New York,
          1984.

[Pré95]   A. Prékopa. *Stochastic Programming*. Kluwer Academic Publishers, Dor-
          drecht, 1995.

[QS62]   J. P. Quirk and R. Saposnik. Admissibility and measurable utility functions.
          Review of Economic Studies, 29:140–146, 1962.

[Rob87]   S. M. Robinson. Local epi-continuity and local optimization. *Mathematical
          Programming*, 37:208–222, 1987.

[Röm03]   W. Römisch. Stability of stochastic programming problems. In
          A. Ruszczyński and A. Shapiro, editors, *Stochastic Programming, Handbooks
          of Operations Research and Management Science*, volume 10, pages 483–554.
          Elsevier, Amsterdam, 2003.

[RS70]   M. Rothschild and J. E. Stiglitz. Increasing risk: I. A definition. *Journal of
          Economic Theory*, 2(3):225–243, 1970.

[RS03]   A. Ruszczyński and A. Shapiro, editors. *Stochastic Programming*, volume 10
          of *Handbooks in Operations Research and Management Science*. Elsevier,
          Amsterdam, 2003.

[RU00]   T. Rockafellar and S. Uryasev. Optimization of conditional value-at-risk. *Jour-
          nal of Risk*, 2(3), 2000.

[Sch95]   R. Schultz. On structure and stability in stochastic programs with random
          technology matrix and complete integer recourse. *Mathematical Program-
          ming*, 70:73–89, 1995.

[Sch96]   R. Schultz. Rates of convergence in stochastic programs with complete integer recourse. *SIAM Journal on Optimization*, 6:1138–1152, 1996.

[Sch00]   R. Schultz. Some aspects of stability in stochastic programming. *Annals of Operations Research*, 100:55–84, 2000.

[She51]   S. Sherman. On a theorem of hardy, littlewood, polya and blackwell. *Proceedings of the National Academy of Sciences*, 37:826–831, 1951.

[SK96]   K. W. Schmitz and G. Koch. *Kraft-Wärme-Kopplung, Anlagenauswahl-Dimensionierung, Wirtschaftlichkeit-Emissionsbilanz*. VDI Verlag, Düsseldorf, 2. edition, 1996.

[SN07]   R. Schultz and F. Neise. Algorithms for mean-risk stochastic integer programs in energy. *Revista Investigación Operational*, 28(1):4–16, 2007.

[ST06]   R. Schultz and S. Tiedemann. Conditional value-at-risk in stochastic programs with mixed-integer recourse. *Mathematical Programming*, 105(2/3):365–386, 2006.

[Tie05]   S. Tiedemann. *Risk Measures with Preselected Tolerance Levels in Two-Stage Stochastic Mixed-Integer Programming*. Cuvillier Verlag, Göttingen, 2005.

[Tin55]   G. Tintner. Stochastic linear programming with applications to agricultural economics. In H. A. Antosiewicz, editor, *Proc. 2nd Symp. Linear Programming, Washington D.C.*, volume 2, pages 197–228, 1955.

[Tob58]   J. E. Tobin. Liquidity preference as behavior towards risk. *Review of Economic Studies*, 67, 1958.

[vdPP63]   C. van de Panne and W. Popp. Minimum cost cattle feed under probabilistic problem constraint. *Management Science*, 9:405–430, 1963.

[VdVW96]   A. W. Van der Vaart and J. A. Wellner. *Weak Convergence and Empirical Processes*. Springer, New York, 1996.

[vNM47]   J. von Neumann and O. Morgenstern. *Theory of Games and Economic Behaviour*. Princeton University Press, Princeton, 1947.

[Wet79]   R. J-B. Wets. *A statistical approach to the solution of stochastic programs with (convex) simple recourse*. Working Paper, Department of Mathematics, University of Kentucky, Lexington, 1979.

[WF78]   G. A. Whitmore and M. C. Findlay, editors. *Stochastic Dominance: An Approach to Decision Making under Risk*. D. C. Heath, Lexington, MA, 1978.

[Whi70]   G. A. Whitmore. Third-degree stochastic dominance. *The American Economic Review*, 60:457–459, 1970.

WWW.VIEWEGTEUBNER.DE

Vieweg+Teubner Research
Ihre wissenschaftliche Arbeit bei uns

Mit unserem neuen Programm Vieweg+Teubner Research möchten wir der Fachwelt herausragende wissenschaftliche Arbeiten aus Technik und Naturwissenschaft präsentieren. Wir veröffentlichen Dissertationen, Habilitationen, Tagungs- und Sammelbände sowie Schriftenreihen – gerne auch Ihre.

Wir bieten Ihnen:

Qualitative Begutachtung und Auswahl der Manuskripte.
Veröffentlichung auch von kleinen Auflagen.
Kurze Produktionszeiten von 6-8 Wochen.
Erstellung Ihrer Pflichtexemplare mit Sonderausstattung gemäß Ihrer Promotionsordnung.
Umfassende Marketing-Aktivitäten und verlegerisches Know-how
aus über 400 Jahren Erfahrung.
Dauerhafte Lieferbarkeit.
Schnelle, weltweite Verfügbarkeit als klassisches Buch und als E-Book (SpringerLink).

Starkes Programm

BAUWESEN | ELEKTROTECHNIK | INFORMATIK UND IT | MASCHINENBAU UND KFZ | MATHEMATIK | NATURWISSENSCHAFTEN

Möchten Sie Autor bei Vieweg+Teubner werden? Kontaktieren Sie uns!
Christel A. Roß | christel.ross@viewegteubner.de | Tel.: +49(0)611.7878-326

VIEWEG+
TEUBNER

TECHNIK BEWEGT.

T0184677

BestMasters

Springer awards „BestMasters" to the best master's theses which have been completed at renowned Universities in Germany, Austria, and Switzerland.

The studies received highest marks and were recommended for publication by supervisors. They address current issues from various fields of research in natural sciences, psychology, technology, and economics.

The series addresses practitioners as well as scientists and, in particular, offers guidance for early stage researchers.

Nikolas Zöller

Optimization of Stochastic Heat Engines in the Underdamped Limit

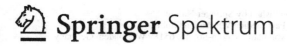

Springer Spektrum

Nikolas Zöller
Potsdam, Germany

BestMasters
ISBN 978-3-658-16349-5   ISBN 978-3-658-16350-1   (eBook)
DOI 10.1007/978-3-658-16350-1

Library of Congress Control Number: 2016956641

Springer Spektrum
© Springer Fachmedien Wiesbaden GmbH 2017
This work is subject to copyright. All rights are reserved by the Publisher, whether the whole or part of the material is concerned, specifically the rights of translation, reprinting, reuse of illustrations, recitation, broadcasting, reproduction on microfilms or in any other physical way, and transmission or information storage and retrieval, electronic adaptation, computer software, or by similar or dissimilar methodology now known or hereafter developed.
The use of general descriptive names, registered names, trademarks, service marks, etc. in this publication does not imply, even in the absence of a specific statement, that such names are exempt from the relevant protective laws and regulations and therefore free for general use.
The publisher, the authors and the editors are safe to assume that the advice and information in this book are believed to be true and accurate at the date of publication. Neither the publisher nor the authors or the editors give a warranty, express or implied, with respect to the material contained herein or for any errors or omissions that may have been made.

Printed on acid-free paper

This Springer Spektrum imprint is published by Springer Nature
The registered company is Springer Fachmedien Wiesbaden GmbH
The registered company address is: Abraham-Lincoln-Str. 46, 65189 Wiesbaden, Germany

Contents

List of Figures

1

Introduction

Classical thermodynamics is the study of general laws governing the transformations of macroscopic systems. These involve the exchange of heat, work and matter between a system and its environment. The foundations of thermodynamics have been established and developed since the nineteenth century by many scientists, its most prominent representatives being Carnot, Mayer, Clausius, Kelvin and Fourier. Cornerstones of the theory were the fundamental publication *Théorie analytique de la chaleur*[1] by Fourier and the *Réflexions sur la puissance motrice du feu et sur les machines propres à développer cette puissance* [2] by Carnot. The microscopic justification of macroscopic thermodynamics at equilibrium was later on provided in the dawn of statistical mechanics, developed by Maxwell, Boltzmann, Gibbs, Onsager, Einstein and others.

The first detailed description of Brownian motion was given by Robert Brown in 1827 when he observed the random motion of pollen in a viscous fluid [3]. Following his work there were many years of speculation until Einstein was the first to make conclusive mathematical predictions for the diffusive behavior of microscopic particles. In his famous 1905 paper [4] he laid the theoretical foundations of thermodynamic fluctuation theory. The same explanations were independently derived by Smoluchowski [5] who together with Perrin was also responsible for much of the later systematic development and experimental verification of Brownian motion theory [6, 7]. Shortly afterwards, Langevin tackled the problem by introducing a fluctuating force within a single particle trajectory approach, thereby providing the first stochastic differential equation [8]. A rigorous mathematical foundation for this technique was not given, until 40 years later Ito [9] formulated his concepts of stochastic differential equations.

For driven Brownian motion, Sekimoto recognized only 20 years ago that two principal notions of classical thermodynamics, namely the exchanged heat and the applied work, could be meaningfully assigned to single particle trajectories [10, 11]. Consequently, he ascribed an analogon of the first law of thermodynamics to the interaction of a microscopic system with its thermal environment and the thermodynamic quantities thereby became fluctuating ones. Sekimoto called this approach stochastic energetics [12].

In this work, we use Sekimoto's framework to study the model of a stochastic heat engine in the high and low damping regime. We focus on the optimization of the power output in the context of finite-time control protocols and emphasize the connection to possible experimental realizations. The optimization of finite-time processes in stochastic thermodynamics has recently drawn a lot of attention [13, 14, 15, 16], whereby calculations were almost exclusively carried out in the overdamped regime. In the context of microscopic heat machines, in particular the efficiency at maximum power was of central interest [17, 18, 19, 20]. These concepts are inherently connected to entropy production and the associated irreversible work. Both notions will be discussed in detail and play an important role in the considerations of this work.

The thesis is structured as follows. In the remainder of this chapter we give an introduction to the single trajectory approach of stochastic motion on microscopic scales and its connection to the evolution of the probability distribution. We thereby discuss the high-friction (overdamped) limit which has dominated calculations done in the literature during the last decades [21]. Subsequently, we explain the concept of stochastic energetics and review classical concepts of heat engines.

In chapter 2 we introduce the concept of a stochastic heat engine and review the theory of optimal protocols in the overdamped limit derived in [22]. Furthermore, we perform a numerical study in which we investigate the transition to the low-friction (underdamped) regime and the associated breakdown of the overdamped theory.

In chapter 3 we describe setups for the experimental realization of a stochastic heat engine in the overdamped and underdamped regime based on colloidal systems and optical cavities. So far, a microscopic heat engine has only been realized in the overdamped regime [23]. Moreover, we extend the theory for overdamped optimal protocols to the case in which the temperature bath difference is created via optical damping in an optical cavity trap.

In chapter 4 we investigate the optimization problem in the underdamped regime. We derive an approximate evolution equation for the probability density and solve this equation analytically for a linear and a piecewise linear driving protocol of a harmonic trapping potential. Subsequently, we numerically optimize the protocol parameters to obtain a maximum power output of the stochastic heat engine.

In chapter 5 we review the theory of the recently derived anomalous entropy production [24] in the presence of temperature gradients and perform a case study for a quadratic temperature profile with possible applications in optical cavities. We thereby compare analytical to numerical results.

Finally, in chapter 6 we summarize our findings and give an outlook for possible continuations of this work.

1.1   Stochastic Dynamics

The main concepts of stochastic dynamics can be introduced by picturing the simple model of a colloidal particle confined in space by some external potential and interacting with its thermal environment. This serves as the acclaimed paradigm in this growing research field [21]. There exist three equivalent but complementary ways to describe such a system, the Langevin equation, the Fokker-Planck equation and the path integral [21]. In this work we will focus on the first two descriptions.

1.1.1 Langevin Equation

A Langevin equation is a Newtonian force equation that describes the trajectory of a particle under the influence of thermal fluctuations due to the interaction with the environment. Its most prominent version is of the form [3]

$$m\ddot{x} = -\gamma\dot{x} - \frac{\partial V(x,t)}{\partial x} + \Gamma(t). \tag{1.1}$$

Here γ is the damping constant that couples to the particle's velocity, $V(x,t)$ is an external potential and $\Gamma(t)$ is a delta-correlated random force also called Langevin force, that describes the coupling to a surrounding heat bath. The thermal fluctuations $\Gamma(t)$ are modeled as Gaussian white noise which has the properties

$$\langle\Gamma(t)\rangle = 0, \quad \langle\Gamma(t)\Gamma(t')\rangle = 2\gamma k_B T\delta(t - t'), \tag{1.2}$$

where k_B is the Boltzmann constant and T denotes the bath temperature. In the physical picture we can think of this stochastic force as a sequence of uncorrelated pulses that give rise to pulses in $\ddot{x}$ and hence jumps in the velocity $\dot{x}$. These pulses emerge due to collisions with other particles in the heat bath. A more general form of a Langevin equation for a N-dimensional variable set $\{\xi\} = \xi_1, \xi_2, ..., \xi_N$ is given by [25]

$$\dot{\xi}_i = h_i(\{\xi\}, t) + g_{ij}(\{\xi\}, t)\Gamma_j(t), \quad \langle\Gamma_j(t)\rangle = 0, \quad \langle\Gamma_i(t)\Gamma_j(t')\rangle = 2\delta_{ij}\delta(t - t'), \tag{1.3}$$

where the Γ_j are again Gaussian random variables with zero mean and a delta correlation in time.

Of particular interest are the Kramers-Moyal coefficients, which are defined as the time derivatives of the n-th order central moments divided by $n!$

$$\begin{aligned}D^{(n)}_{i_1\cdots j_n}(\{x\}, t) &= \frac{1}{n!}\frac{d}{dt}M_{i_1\cdots j_n}(t) \\ &= \frac{1}{n!}\lim_{\tau\to 0}\frac{1}{\tau}\langle[\xi_{i_1}(t + \tau) - x_{i_1}]...[\xi_{j_n}(t + \tau) - x_{j_n}]\rangle\Big|_{\xi_i(t)=x_i,...,\xi_{j_n}(t)=x_{j_n}}.\end{aligned} \tag{1.4}$$

Because of the properties of the Gaussian white noise, it can be shown [25] that for a process described by equation (1.3) these coefficients vanish for $n > 2$ and the first two are given by

$$D^{(1)}_i(\{x\}, t) = h_i(\{x\}, t) + g_{kj}(\{x\}, t)\frac{\partial}{\partial x_k}g_{ij}(\{x\}, t) \tag{1.5}$$

$$D^{(2)}_{ij}(\{x\}, t) = g_{ik}(\{x\}, t)g_{jk}(\{x\}, t), \tag{1.6}$$

where here and in the following we use Einstein's summation convention for repeated indices. $D^{(1)}_i$ is called the drift and $D^{(2)}_{ij}$ the diffusion coefficient. When known, they are sufficient to set up an evolution equation for the probability distribution, which is explained in the subsequent section. Essentially, the drift and diffusion coefficients provide the link between the two descriptions of a stochastic process via a Langevin or a Fokker-Planck equation. Note that while the transition from Langevin to Fokker-Planck equation is always unique, the reverse is only true in one dimension [26, 27].

1.1.2   Fokker-Planck Equation

Although in principle we can use equation (1.1) to calculate expectation values, it is in many cases more convenient to work with an evolution equation for the probability density. An equation of this kind was for the first time derived by Fokker [28] and Planck [29]. For notational simplicity we restrict ourselves in the following derivation to the one-dimensional case but generalization is straightforward [25].

The probability density $W(x, t + \tau)$ is connected to the probability density at time t via the transition probability $\mathcal{P}$

$$W(x, t + \tau) = \int \mathcal{P}(x, t + \tau | x', t) W(x', t) dx', \tag{1.7}$$

where the bar | denotes a conditional probability, i.e. the probability that the particle is found at position x at time $t + \tau$ under the condition that it was at position x' at time t. Assuming that we know all the moments M_n we can construct the characteristic function

$$C(u, x', t, \tau) = \int_{-\infty}^{\infty} e^{iu(x-x')} \mathcal{P}(x, t + \tau | x', t) dx = 1 + \sum_{n=1}^{\infty} (iu)^n M_n(x', t, \tau)/n!. \tag{1.8}$$

Performing a Fourier transform we obtain

$$\mathcal{P}(x, t + \tau | x', t) = \frac{1}{2\pi} \int_{-\infty}^{\infty} e^{-iu(x-x')} C(u, x', t, \tau) du$$

$$= \frac{1}{2\pi} \int_{-\infty}^{\infty} e^{-iu(x-x')} \left[1 + \sum_{n=1}^{\infty} (iu)^n M_n(x', t, \tau)/n! \right] du. \tag{1.9}$$

Since $n \geq 0$ we can use the following properties of the Dirac delta distribution

$$\frac{1}{2\pi} \int_{-\infty}^{\infty} (iu)^n e^{-iu(x-x')} du = \left(-\frac{\partial}{\partial x} \right)^n \delta(x - x'), \tag{1.10}$$

$$\delta(x - x') f(x') = \delta(x - x') f(x), \tag{1.11}$$

to arrive at

$$\mathcal{P}(x, t + \tau | x', t) = \left[1 + \sum_{n=1}^{\infty} \frac{1}{n!} \left(-\frac{\partial}{\partial x} \right)^n M_n(x, t, \tau) \right] \delta(x - x'). \tag{1.12}$$

Inserting (1.12) into (1.7) leads to

$$W(x, t + \tau) - W(x, t) = \frac{\partial W(x, t)}{\partial t} \tau + O(\tau^2) \tag{1.13}$$

$$= \sum_{n=1}^{\infty} \frac{1}{n!} \left(-\frac{\partial}{\partial x} \right)^n M_n(x, t, \tau) W(x, t). \tag{1.14}$$

We now perform a Taylor expansion of the moments with respect to τ:

$$M_n(x, t, \tau)/n! = D^{(n)}(x, t)\tau + O(\tau^2), \tag{1.15}$$

where terms of order τ^0 are absent due to the initial condition at $\tau = 0$:

$$\mathcal{P}(x, t|x', t) = \delta(x - x').\tag{1.16}$$

Taking only linear terms in τ we arrive at the Kramers-Moyal forward expansion

$$\frac{\partial W(x,t)}{\partial t} = \sum_{n=1}^{\infty} \left(-\frac{\partial}{\partial x}\right)^n D^{(n)}(x,t)W(x,t) = L_{KM}^\dagger W.\tag{1.17}$$

Since the transition probability $\mathcal{P}(x, t|x', t')$ is the distribution $W(x, t)$ for the special initial condition $W(x, t') = \delta(x - x')$, it itself obeys equation (1.17), i.e

$$\frac{\partial \mathcal{P}(x,t|x',t')}{\partial t} = \sum_{n=1}^{\infty} \left(-\frac{\partial}{\partial x}\right)^n D^{(n)}(x,t)\mathcal{P}(x,t|x',t') = L_{KM}^\dagger \mathcal{P}(x,t|x',t').\tag{1.18}$$

Note that in (1.18) the occurring differential operators are with respect to x and t, and not with respect to x' and the earlier time t'. This is the reason why the expansion is called *forward* expansion and why we denoted the operator $L_{KM}^\dagger$ with a dagger as the adjoint. A backward expansion can equally be performed (see [25]) and leads to the equation

$$\frac{\partial \mathcal{P}(x,t|x',t')}{\partial t'} = -\sum_{n=1}^{\infty} D^{(n)}(x,t)\left(\frac{\partial}{\partial x'}\right)^n \mathcal{P}(x,t|x',t') = -L_{KM}\mathcal{P}(x,t|x',t').\tag{1.19}$$

Equations (1.18) and (1.19) are equivalent since they lead to the same solution for the probability distribution $W(x, t)$.

The multi-variable generalization of equation (1.17) is straightforward and reads

$$\frac{\partial W(\{x\},t)}{\partial t} = \sum_{n=1}^{\infty} \frac{(-\partial)^n}{\partial x_{j_1} ... \partial x_{j_N}} D^{(n)}_{j_1,...,j_n}(\{x\},t)W(\{x\},t) = L_{KM}^\dagger W(\{x\},t).\tag{1.20}$$

If the series in equation (1.20) is truncated after $n = 2$ or the higher coefficients vanish, which they do for a Langevin Equation with Gaussian white noise [25], we arrive at the Fokker-Planck equation

$$\frac{\partial W(\{x\},t)}{\partial t} = L_{FP}^\dagger W(\{x\},t), \quad L_{FP}^\dagger = -\frac{\partial}{\partial x_i}D_i^{(1)}(\{x\},t) + \frac{\partial^2}{\partial x_i \partial x_j}D_{ij}^{(2)}(\{x\},t).\tag{1.21}$$

In particular the Fokker-Planck equation corresponding to (1.1) is called Klein-Kramers equation[30] and is given by

$$\frac{\partial \mathcal{P}(x,v,t)}{\partial t} = \left[-v\frac{\partial}{\partial x} + \frac{1}{m}\frac{\partial}{\partial v}\left(\frac{\partial V}{\partial x} + \gamma v\right) + \frac{\gamma k_B T}{m^2}\frac{\partial^2}{\partial v^2}\right]\mathcal{P}(x,v,t).\tag{1.22}$$

1.1.3 Wiener Process

The Wiener process[1] is a stochastic process in one variable characterized by a vanishing drift coefficient $D^{(1)} = 0$ and a diffusion coefficient of $D^{(2)} = 1/2$. We introduce it here not

---

[1]named in honor of Norbert Wiener, an American mathematician and philosopher (1894–1964) [31].

only as the simplest possible example of a diffusion process, but also because it provides a basis for a rigorous mathematical description of stochastic differential equations[2]. The Fokker-Planck equation governing the evolution of the probability distribution reads

$$\frac{\partial}{\partial t}\mathcal{P}(w,t|w_0,t_0) = \frac{1}{2}\frac{\partial^2}{\partial w^2}\mathcal{P}(w,t|w_0,t_0),$$ (1.23)

with the initial condition

$$\mathcal{P}(w,t_0|w_0,t_0) = \delta(w-w_0)$$ (1.24)

and w being the random variable. Defining the characteristic function

$$\phi(k,t) = \int \mathcal{P}(w,t|w_0,t_0)e^{ikw}dw,$$ (1.25)

we take its time derivative, use equation (1.23) and perform a partial integration to obtain

$$\frac{\partial\phi}{\partial t} = -\frac{1}{2}k^2\phi.$$ (1.26)

This equation is easily solved and its solution reads

$$\phi(k,t) = \exp\left[ikw_0 - \frac{1}{2}k^2(t-t_0)\right],$$ (1.27)

where we have used the initial condition in equation (1.24). An inverse Fourier transform then provides the solution for the conditional probability density

$$\mathcal{P}(w,t|w_0,t_0) = [2\pi(t-t_0)]^{-1/2}\exp[-(w-w_0)^2/2(t-t_0)],$$ (1.28)

which is a Gaussian centered around w_0 with width $t - t_0$. The sample paths of the stochastic process, which we denote by $W(t)$, are continuous but not differentiable. This is easily seen by considering the probability for the absolute value of the derivative being greater than some constant k:

$$\lim_{h\to 0}\text{Prob}\{|[W(t+h)-W(t)]/h| > k\} = \lim_{h\to 0} 2\int_{kh}^{\infty} dw(2\pi h)^{-1/2}\exp(-w^2/2h) = 1,$$ (1.29)

where we have used the solution of the Wiener process from equation (1.28). The result is independent of the constant k, which means that the derivative at any point is almost certainly infinite. Hence, the sample paths are not differentiable.

With the initial condition $w_0 = 0$ the Langevin equation belonging to the Wiener process is given by

$$\frac{dW}{dt} = \Gamma(t).$$ (1.30)

But we have seen, that $W(t)$ is not differentiable and therefore equation (1.30) is mathematically not well defined. On the other hand we can write the formal solution of $W(t)$ as the integral

$$W(t) = \int_0^t \Gamma(t')dt$$ (1.31)

which can be interpreted consistently.

---

[2]This plays an important role in the correct description of multiplicative noise, which will be of relevance in chapter 5.

1.1.4 Stochastic Differential Equations (SDE)

Following the argument from section 1.1.3, the Langevin equation as introduced in equation (1.3) is mathematically not well defined. If the prefactors g_{ij} are independent of the random variables, which means that we do not have multiplicative noise, this does not lead to further complications. But if the $g_{ij}(\xi(t), t)$ depend on the random variables we have to reformulate the Langevin equation as an integral equation and specify at which point the stochastic integrals are evaluated [32]. Equation (1.3) is then rewritten as

$$\xi_i - \xi_i(0) = \int_0^t h_i(\{\xi(s)\}, s)ds + \int_0^t g_{ij}(\{\xi(s)\}, s)\Gamma_j(s)ds. \tag{1.32}$$

We now make the replacement

$$dW_j(t) \equiv W_j(t + dt) - W_j(t) = \Gamma_j dt \tag{1.33}$$

which directly follows from the interpretation of the integral of $\Gamma(t)$ as the Wiener process (see equation (1.31)). Equation (1.32) is then written as

$$\xi_i - \xi_i(0) = \int_0^t h_i(\{\xi(s)\}, s)ds + \int_0^t g_{ij}(\{\xi(s)\}, s)dW(s). \tag{1.34}$$

This equation is still meaningless, unless we specify how exactly the stochastic integral $\int_0^t g_{ij}(\{\xi(s)\}, s)dW(s)$ is evaluated [32].

Stochastic Integration

A stochastic integral of the form $\int_{t_0}^t G(t')dW(t')$ is defined as a limit of the partial sums

$$\lim_{n \to \infty} S_n = \lim_{n \to \infty} \sum_{i=1}^n G(\tau_i)[W(t_i) - W(t_{i-1})], \tag{1.35}$$

where the partitioning of the time interval follows

$$t_0 \leq t_1 ... \leq t_{n-1} \leq t, \text{ with } t_{i-1} \leq \tau_i \leq t_i. \tag{1.36}$$

Since $W(t)$ is not differentiable, S_n generally depends on the particular choice of the intermediate points τ_i at which $G(\tau_i)$ is evaluated. Two integrals have historically earned the most attention in the field of stochastic thermodynamics [32]. The first is the Ito stochastic integral which is defined as

$$\int_{t0}^t G(t') \cdot dW(t') = \lim_{n \to \infty} \sum_{i=1}^n G(t_{i-1})[W(t_i) - W(t_{i-1})], \tag{1.37}$$

and corresponds to evaluating G at the beginning of the time interval. This integration rule is denoted by the · in the product under the integral. The second is the Stratonovich integral defined as

$$\int_{t_0}^t G(t') \circ dW(t') = \lim_{n \to \infty} \sum_{i=1}^n \frac{G(t_i) + G(t_{t-1})}{2}[W(t_i) - W(t_{i-1})], \tag{1.38}$$

where the function G is averaged over the beginning and the end point of the time interval and the rule for evaluation is denoted by the $\circ$ sign in the product under the integral.

Ito and Stratonovich SDE

These two stochastic integrals lead to different interpretations of the integral Langevin equation (1.34). As stated before, the Langevin equation is meaningless unless a rule of how to interpret the stochastic integral is provided. The reason may be understood as follows. The stochastic force gives rise to a jump in ξ, but since the prefactor of this force itself depends on ξ it is not clear where $g(\xi, t)$ should be evaluated and therefore is undetermined. The Ito convention assigns a meaning to equation (1.34) by demanding that the values of ξ and t in $g(\xi, t)$ should be taken just before the pulse of the stochastic force. This defines the Ito SDE as

$$d\xi(t) = h[\xi(t), t]dt + g[\xi(t), t] \cdot dW(t), \tag{1.39}$$

where the dot denotes that subsequent integration is performed with respect to the Ito stochastic integral. Its solution can be shown to be a Markov process and the properties of the Ito stochastic integral are often useful for formal mathematical proofs [6]. A serious difficulty on the other hand is that the usual rules of differential calculus cannot be applied without caution, since $dW^2 \propto dt$ and so higher order terms have to be taken into account. This is important especially when one wants to perform non-linear coordinate transformations. One therefore has to learn and apply the rules of Ito-Calculus, for an introduction we point out reference [6].

The Stratonovich differential equation is defined as

$$d\xi(t) = h[\xi(t), t]dt + g[\xi(t), t] \circ dW(t) \tag{1.40}$$

and the $\circ$ denotes that subsequent integration is performed with respect to the Stratonovich stochastic integral. For the Stratonovich SDE the rules for change of variables are exactly the same as in ordinary calculus.

Note that equations 1.39 and 1.40 are not the same and may lead to different results for mean values and probability distributions. But this does not mean that one or the other is correct or incorrect. With the same function $h[\xi(t), t]$ they simply describe two different stochastic processes. The same stochastic process can either be described by an Ito or a Stratonovich SDE and the two merely mathematical representations are connected via the transformation [32]:

Stratonovich: $$d\xi(t) = h dt + g \circ dW \tag{1.41}$$

Ito: $$d\xi(t) = \left(h + \frac{g}{2} \frac{\partial g}{\partial \xi} \right) dt + g \cdot dW. \tag{1.42}$$

Note that unless explicitly stated otherwise we generally adopt the Stratonovich representation in this work and drop the $\circ$ for simplicity of denotation.

1.1.5 High-Friction Limit

If the friction constant in Eq. (1.1) is very large, the damping term outweighs the influence of inertia and as a consequence the mass term may be neglected. Equation (1.1) then simplifies to

$$\gamma \dot{x} = -\frac{\partial V(x,t)}{\partial x} + \Gamma(t). \tag{1.43}$$

The validity of this regime is restricted by the condition $\gamma t/m \gg 1$ [12] which in most cases of stochastic motion observed in nature is fulfilled. The overdamped limit has therefore rightfully been the paradigm in which the calculations of most publications over the last decades have been carried out [21]. Nevertheless, it is important to note that for the overdamped limit to be justified not only a high friction to mass ratio is essential, but also the timescale in which one is interested. In particular at very short timescales the overdamped approximation will always fail, since the condition $\gamma t/m \gg 1$ is not fulfilled. On the other hand for very large times and a slowly changing or static potential the overdamped limit might still be a good approximation even for lower friction to mass ratios. The timescale in which one is interested is therefore of uttermost importance.

Calculating the drift and diffusion coefficients from equation (1.43) and following the formalism of section 1.1.2 one can derive an evolution equation of the probability in position space only

$$\frac{\partial \mathcal{P}(x,t)}{\partial t} = \frac{\partial}{\partial x} \frac{1}{\gamma} \left(\frac{\partial V}{\partial x} + k_B T \frac{\partial}{\partial x} \right) \mathcal{P}(x,t), \tag{1.44}$$

which was first discovered by Smoluchowski [5]. The Smoluchowski Equation can also be derived by an inverse friction expansion starting from the Klein-Kramers Equation (1.22) which for the first time was rigorously done by Wilemski [33]. In appendix A.1 we show an approach [24] that emphasizes the separation of timescales, for an introduction to the technique see reference [34].

1.2 Stochastic Energetics

The term stochastic energetics was introduced by Sekimoto in his seminal work on the links between macroscopic thermodynamics and microscopic fluctuation theory [10, 11]. He realized that central concepts of classical thermodynamics like work and heat could be meaningfully defined on the the level of single particle trajectories. We here give a short introduction to his main ideas, whereas an extensive treatment can be found in Sekimoto's book [12].

1.2.1 Thermodynamics of Single Trajectories

At the heart of macroscopic thermodynamics stands the first law, which essentially is a conservation law for the energy of the system and its environment

$$dU = d'Q + d'W. \tag{1.45}$$

Here $d'Q$ denotes the heat transferred to the system, $d'W$ is the work performed on the system and dU is the total change of energy of the system, whereby d' just denotes that the heat and work increments are not total differentials. The signs are just convention in which direction heat and work flows are defined positive, but it is important to stick to

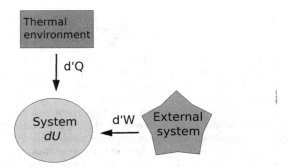

Figure 1.1: Sketch of the energy balance of a system, for example a particle, with its thermal environment and an external system.

one such convention. A sketch of the interactions of the system with its environment is given in figure 1.1. The objective is now to find a way to endow single particle trajectories with an energy balance equation like the first law of thermodynamics. The Fokker-Planck equation is not a reasonable description for this purpose, because it deals with an ensemble of stochastic processes, but not an individual one. We therefore start with the Langevin equation (1.1)

$$m\ddot{x} = -\gamma\dot{x} - \frac{\partial V(x,t)}{\partial x} + \Gamma(t), \quad \langle\Gamma(t)\rangle = 0, \quad \langle\Gamma(t)\Gamma(t')\rangle = 2\gamma k_B T\delta(t - t'). \qquad (1.46)$$

The direct interaction of the particle with the bath is modeled through a velocity dependent friction term and the stochastic force Γ. Sekimoto therefore suggested to define heat as the energy transfer via the relation

$$d'Q \equiv \left(-\gamma\frac{dx}{dt} + \Gamma(t)\right) dx(t), \qquad (1.47)$$

where $dx(t)$ is the evolution of x over a time interval dt. This definition, combined with the identities for kinetic and potential energy

$$\frac{dp}{dt}dx(t) = d\left(\frac{p^2}{2m}\right), \qquad (1.48)$$

$$\frac{\partial V}{\partial x}dx(t) = dV(x(t), \lambda(t)) - \frac{\partial V}{\partial\lambda}d\lambda, \qquad (1.49)$$

gives

$$dU = d'Q + \frac{\partial V}{\partial\lambda}d\lambda. \qquad (1.50)$$

Note that the time dependence of the potential is entirely encoded in the control parameter $\lambda(t)$. Interpreting equation (1.50) as a trajectory based first law of thermodynamics analogously to equation (1.45), we identify the work as

$$d'W \equiv \frac{\partial V}{\partial\lambda}d\lambda, \qquad (1.51)$$

where λ is the control parameter of the external potential and inhibits its time dependence. Note that unlike in classical mechanics the definition of work here relies on a generalized force $\partial V/\partial\lambda$ which refers to the control parameter λ and not to the position variable x.

Having expressed the first law along an individual trajectory, we want to define entropy along similar terms. The common definition of a non-equilibrium Gibbs entropy

$$S = -k_B \int \mathcal{P}(x,v,t) \ln \mathcal{P}(x,v,t) dx dv. \tag{1.52}$$

is essentially an ensemble quantity, where $\mathcal{P}$ is the solution of the Fokker-Planck equation. Interpreting it as an ensemble average suggests to define a trajectory-dependent entropy for a single particle [35] as

$$s = -k_B \ln \mathcal{P}(x(t), v(t), t), \tag{1.53}$$

where $\mathcal{P}$ is evaluated along the trajectory $x(t)$ and $v(t)$. Note that the stochastic entropy therefore not only depends on the individual trajectory, but also on the ensemble from which it is taken. For the calculation of the total entropy, also the heat dissipated into the environment has to be considered. Using equations (1.46) and (1.47) the entropy produced in the environment becomes

$$s_{env} = -\int_{t_0}^{t} \frac{1}{T} \frac{dQ}{dt} dt = -\int_{t_0}^{t} \frac{1}{T} \left(\frac{\partial V}{\partial x} v + mv\dot{v} \right) dt, \tag{1.54}$$

and the total entropy change of particle and environment is

$$s_{tot} = s_p + s_{env}. \tag{1.55}$$

The great advantage of assigning thermodynamic quantities to single trajectories is, that we can now calculate these in single particle experiments by simply measuring the trajectories. This is of importance, especially in the light of recent advances in experimental trapping techniques [36, 23, 37, 38, 39]. In fact, the validity of the first law-like balance between applied work, exchanged heat, and internal energy on the level of a single trajectory has been experimentally demonstrated for the first time by Blickle *et. al.* [40] in 2006. A picture of their results is given in 1.2.

By repeated or longtime measurements it is also possible to obtain the according ensemble averages Besides, computer simulations of Langevin dynamics enable us to calculate work and heat numerically and allow for verification of experimental results.

1.2.2 Ensemble Averages

In the last subsection we derived expressions for work and heat along stochastic trajectories. Thereby these quantities became stochastic themselves and their values differ from one realization to another. In order to make experimentally verifiable predictions, we are interested in the calculation of ensemble averages. In Eq. (1.47) we derived an equation for the heat increment transferred to the system during the time interval dt. To calculate

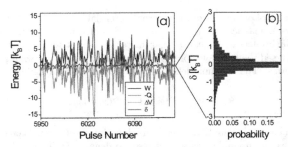

Figure 1.2: Experimental verification by Blickle *et. al.* of the first law-like balance between applied work W, exchanged heat Q, and internal energy ΔV on the level of a single trajectory. (a) Thermodynamic quantities measured along the trajectories of a colloidal particle trapped in an optical tweezer. (b) Distribution histogram of $\delta = W - Q - \Delta V$, the experimentally observed deviation from the first law. Image taken from [40].

the ensemble average of this quantity, it is advantageous to translate to the Ito-formulation (for a derivation see [12])

$$d'Q = -\frac{2\gamma}{m}\left(\frac{p^2}{2m} - \frac{k_B T}{2}\right)dt + \sqrt{2\gamma k_B T}\frac{p}{m} \cdot dW. \qquad (1.56)$$

Taking the average, the last term vanishes due to the non-anticipating character of the prefactor of dW and we arrive at

$$\langle d'Q \rangle = -\frac{2\gamma}{m}\left(\frac{\langle p^2 \rangle}{2m} - \frac{k_B T}{2}\right)dt. \qquad (1.57)$$

This result indicates, that independent of the form of the potential $V(x, \lambda(t))$ the system exchanges heat with its thermal environment via the kinetic degrees of freedom.

In the overdamped limit the velocity is always assumed to be Maxwell-distributed, which for the average kinetic energy means $\langle p^2 \rangle/2m = k_B T/2$. This assumption substituted in equation (1.57), however leads to a zero heat transfer $\langle d'Q \rangle = 0$ for any process, which obviously cannot be correct. The definition of heat $d'Q$ in the overdamped limit has therefore to be reformulated [12] and the result turns out to be

$$\langle d'Q \rangle = -\frac{1}{\gamma}\left[\left\langle\left(\frac{\partial V}{\partial x}\right)^2\right\rangle - \left\langle\frac{\partial^2 V}{\partial x^2}\right\rangle\right]dt. \qquad (1.58)$$

The above results can be presented in a common form, using the probability density $\mathcal{P}$ which is determined by the Fokker-Planck equation. The energy balance for averaged quantities $\langle dU \rangle = \langle d'Q \rangle + \langle d'W \rangle$ leads to

$$\frac{\langle d'Q \rangle}{dt} = \frac{d}{dt}\int U\mathcal{P}d\Gamma - \frac{d\lambda}{dt}\int\frac{\partial U}{\partial\lambda}\mathcal{P}d\Gamma = \int U\frac{\partial\mathcal{P}}{\partial t}d\Gamma, \qquad (1.59)$$

where we have used the fact that $\partial U/\partial \lambda = \partial V/\partial \lambda$ and $d\Gamma = dpdx$ (overdamped $d\Gamma = dx$) denotes the infinitesimal phase space volume. The ensemble averaged heat flux can then be expressed quite generally in terms of probability fluxes using the continuity equation $\partial \mathcal{P}/\partial t = -\nabla \mathbf{J}$. Substituting this into Eq. (1.59) and performing a partial integration whereby the boundary parts vanish, we obtain

$$\frac{\langle d'Q \rangle}{dt} = \int \left[\frac{\partial U}{\partial x} J_x + \frac{\partial U}{\partial p} J_p \right] dxdp \quad \text{(underdamped)}, \tag{1.60}$$

$$\frac{\langle d'Q \rangle}{dt} = \int \frac{\partial U}{\partial x} J_x dx \quad \text{(overdamped)}, \tag{1.61}$$

where the fluxes J_x and J_p can be extracted from the according Fokker-Planck equation. These expressions were solely in the context of the Fokker-Planck equation first derived by Spohn and Lebowitz [41].

The average work is given by the ensemble averages of the single trajectory definition (1.51)

$$\langle \Delta W \rangle = \int_{t_i}^{t_f} \int \mathcal{P} \frac{\partial V}{\partial \lambda} \dot{\lambda} d\Gamma dt. \tag{1.62}$$

The average total entropy production rate can then be written as the time derivative of the sum of the system entropy and the entropy dissipated into the thermal environment due to the heat flux

$$\frac{dS_{tot}}{dt} = \frac{dS}{dt} - \frac{1}{T} \frac{\langle d'Q \rangle}{dt}. \tag{1.63}$$

Using equations (1.52), (1.60) (overdamped (1.61)) and the according Fokker-Planck equation (1.22) (overdamped (1.44)) in equation (1.65), we can reformulate the expression for the average total entropy production, which becomes

$$\frac{dS_{tot}}{dt} = \int \frac{1}{\mathcal{P}} \frac{\gamma}{T} \left[\frac{p}{m} \mathcal{P} + k_B T \frac{\partial \mathcal{P}}{\partial p} \right]^2 dxdp \quad \text{(underdamped)}, \tag{1.64}$$

$$\frac{dS_{tot}}{dt} = \int \frac{\gamma}{T} \frac{J_x^2}{\mathcal{P}} dx \quad \text{(overdamped)}. \tag{1.65}$$

We see that the right hand sides for both cases, full dynamics and overdamped limit, are of quadratic form and the average total entropy production is therefore positive, i.e $S_{tot} > 0$. This essentially is a generalization of the second law of macroscopic thermodynamics. However, note that although the mean values are positive, entropy production is a fluctuating quantity and along single trajectories there is a non-vanishing probability for it to become negative. These so called "violations of the second law" have created much interest during the last decades of research [42, 43].

1.3 Classical Heat Engines

In everyday life we are surrounded by heat engines, refrigerators and other energy transforming devices which are a cornerstone of current technology and infrastructure. The

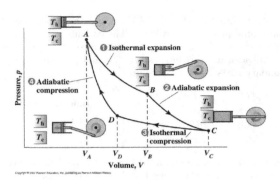

Figure 1.3: Working principle of a Carnot engine. Image taken from Ref. [44].

study of these devices and their related processes has been essential to the past development and will be equally important in the future. In the past the study and construction of such devices has been restricted to the macroscopic regime, but with the emergence of new experimental techniques that allow for the manipulation and design of microscopic processes the possibility to move to smaller scales arose. In a first step we will review two of the most important concepts of heat engines in the macroscopic description which shaped our way of thinking, the Carnot engine and the Curzon-Ahlborn engine. We will then introduce a model for a stochastic heat engine and investigate optimizations and limitations.

1.3.1 Carnot Engine

The Carnot engine, introduced by Carnot in 1824 [2] and depicted in figure 1.3, is a hypothetical engine that operates on the reversible Carnot cycle. It is composed of four basic steps. 1. Reversible isothermal expansion of the gas when coupled to the hot bath. 2. Isentropic expansion in which the system is thermally insulated and no heat is transferred. 3. Reversible isothermal compression when coupled to the cold bath. 4. Isentropic compression in which the system is thermally insulated and again no heat is transferred. The efficiency which is defined as the work output divided by the heat uptake is then determined to be

$$\eta_C = 1 - \frac{T_c}{T_h}. \tag{1.66}$$

There are two main problems concerning the applicability of the Carnot engine to real engines. First, to be reversible the engine is operated at quasi-equilibrium which requires infinite cycle times. Second, as a consequence the power output of the system, defined as work output divided by the cycle time, becomes zero. An attempt to tackle these problems was given by Curzon and Ahlborn [45] as described in the following subsection. A compact review on recent advances on the problem of efficiency of heat engines operating at maximum power can be found in [17]

1.3.2 Curzon-Ahlborn Engine

The heat engine considered by Curzon and Ahlborn operates on a Carnot-like cycle as described in the previous section. The main difference is the fact, that the reversibility is lost by assuming finite cycle times; instead of reversible we now deal with endoreversible processes. Three core assumptions were formulated by Curzon and Ahlborn to derive an efficiency at maximum power. The first states that the working substance and the reservoir operate at different temperatures. This leads to a heat flux, which is assumed to be proportional to the temperature difference and the time that the coupling to the bath is maintained

$$Q_1 = \alpha(T_h - T_{he})t_1,$$
$$Q_3 = \beta(T_{ce} - T_c)t_3. \tag{1.67}$$

Secondly, the total time for the duration of the cyclic process only depends on the isothermal processes. The switching between the baths is assumed to be instantaneous

$$t_{tot} = c(t_1 + t_3). \tag{1.68}$$

Here c is a proportionality constant, while t_1 and t_3 are the coupling times to the hot and cold bath respectively. With these two assumptions an expression for the power $P = -W/t_{tot}$ can be derived

$$P = \frac{T_{he} - T_{ce}}{c[T_{he}/\alpha(T_h - T_{he}) + T_{ce}/\beta(T_{ce} - T_c)]}, \tag{1.69}$$

where the total time has been expressed through the effective temperature of the working substance. Maximization of the power in equation (1.69) leads to a relation between effective and bath temperatures

$$\frac{T_{ce}}{T_{he}} = \sqrt{\frac{T_c}{T_h}}. \tag{1.70}$$

The third assumption states, that we can map the engine to a Carnot engine operating at effective temperatures. We thereby arrive at an expression for the efficiency at maximum power

$$\eta_{CA} = 1 - \sqrt{\frac{T_c}{T_h}}. \tag{1.71}$$

For a long time this expression had been thought to provide a universal upper bound for the efficiency at maximum power of Carnot-like heat engines and it was only shown in 2008 by Seifert and Schmiedl [22] that in principle higher efficiencies are possible.

2

Stochastic Heat Engine - Overdamped Regime

We now move to smaller scales where microscopic fluctuations have to be taken into account. Particle motion of this kind is naturally described within the framework of Langevin dynamics outlined in section 1.1.1 and the probability distributions in phase space are governed by the formalism of the Fokker-Planck equation 1.1.2. The formulation of stochastic energetics by Sekimoto [10, 11] allows for the definition of thermodynamic quantities such as work and heat along single stochastic trajectories, where the respective ensemble quantities are obtained after averaging (see section 1.2).

Brownian motors are mostly driven by time-dependent potentials, or chemical potential differences. Thereby transport phenomena often are of central interest. An extensive review by Reimann on Brownian motors that mainly focuses on ratchets can be found in [46]. To extract work from ratchet heat engines temperature gradients on small length scales are necessary. These are hard to realize experimentally and, apart from that, do not lead to heat engines of the Carnot type in the macroscopic limit.

We here focus on a stochastic heat engine that operates at two different temperature baths which are spatially homogeneous. A model for such an engine operating in the overdamped regime has been introduced by Schmiedl and Seifert [22] in 2008 and we will follow their approach.

2.1 Carnot-type Model of a Stochastic Heat Engine

In the model, which is sketched in figure 2.1, Brownian particles are confined within a trapping potential that can be controlled externally by a time-dependent control parameter $\lambda(t)$. Similarly to a macroscopic Carnot-like engine (see section 1.3.1) the engine is run through periodic cycles that consist of four steps, steps 1 and 3 being isothermal processes while 2 and 4 are adiabatic transitions. During the isothermal steps the system is coupled to a constant temperature bath. The trap is opened while coupled to the warm bath, letting the particle distribution expand, and closed while coupled to the cold bath, confining the particles again. The adiabatic steps are idealized as sudden jumps of the potential while uncoupling from one temperature bath and coupling to the other. The mean work

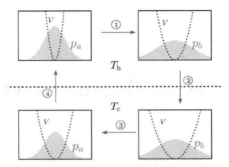

Figure 2.1: Scheme for the cyclic process of a stochastic heat engine introduced by Seifert and Schmiedl [22]. The dotted-curve represents a trapping potential $V(x,t)$ in which the particle distribution is confined. Steps 1 and 3 are isothermal processes interrupted by adiabatic steps 2 and 4 in which the switching between the heat baths is performed. Image taken from Ref. [22].

performed in such a transition is given as

$$W = \Delta E = \int dx P(x,t)[V(x,t+0) - V(x,t-0)]. \tag{2.1}$$

A cyclic process imposes periodic conditions on the particle distribution and on the potential

$$\mathcal{P}(x, t_1 + t_3) = \mathcal{P}(x, 0), \quad V(x, t_1 + t_3) = V(x, 0). \tag{2.2}$$

Furthermore, the energy difference and the entropy difference for a complete cycle have to add up to zero

$$\sum_{i=1}^{4} \Delta E^{(i)} = \sum_{i=1}^{4} \Delta S^{(i)} = 0, \tag{2.3}$$

whereby S is the system's entropy as defined in equation (1.52). Since the particle distribution is not changed during the instantaneous adiabatic steps, $\Delta S^{(2)} = \Delta S^{(4)} = 0$, we have $\Delta S^{(1)} = -\Delta S^{(3)} = \Delta S$ and the total work can be written as

$$W = \sum_{i=1}^{4} W^{(i)} = \sum_{i=1}^{4} W_{\text{irr}}^{(i)} - \sum_{i=1}^{4} T_i \Delta S^{(i)} + \sum_{i=1}^{4} \Delta E^{(i)} \tag{2.4}$$

$$= W_{\text{irr}}^{(1)} + W_{\text{irr}}^{(3)} - (T_h - T_c)\Delta S. \tag{2.5}$$

The irreversible work W_{irr} is defined as the work lost due to the total irreversible entropy production ΔS_{tot} of both, system and the thermal environment combined. Using equation (1.65) it is in the overdamped limit given by

$$W_{irr} = \int_{t_i}^{t_f} dt T \frac{dS_{tot}}{dt} = \int_{t_i}^{t_f} dt \int \frac{\gamma J_x^2}{\mathcal{P}} dx. \tag{2.6}$$

2.2 Optimal Protocols in the Overdamped Limit for an Harmonic Trapping Potential

In this section we will work with dimensionless units. In order to reobtain actual values one has to multiply with the corresponding units given in equation (A.2). We assume the trapping potential to be harmonic

$$V(x,t) = \lambda(t)x^2/2. \tag{2.7}$$

This case is important, because most experimental trapping techniques can be approximated by such a potential [47]. The objective is to derive optimal protocols for the control parameter λ. By optimal protocol we refer to a protocol for which the irreversible work during the process is minimized.

We proceed by splitting the heat cycle into parts at constant temperature. In the overdamped regime the time evolution of $\mathcal{P}(x,t)$ is then governed by the Smoluchowski equation (1.44) which for the harmonic potential (2.7) and constant γ and T reads

$$\frac{\partial \mathcal{P}}{\partial t} = \frac{\lambda}{\gamma} \frac{\partial}{\partial x}(x\mathcal{P}) + \frac{T}{\gamma} \frac{\partial^2 \mathcal{P}}{\partial x^2}. \tag{2.8}$$

Multiplying equation (2.8) by x^2 and performing partial integrations, we arrive at a differential equation, which connects the control parameter λ and the second moment $\omega \equiv \langle x^2 \rangle$:

$$\dot{\omega} = \frac{2}{\gamma}(T - \lambda\omega). \tag{2.9}$$

Using equations (1.62) and (2.9) the mean work can be cast into a local functional of the variance ω only:

$$W[\lambda(t)] = \int_{t_i}^{t_f} dt \dot{\lambda} \frac{\omega}{2} = \frac{\gamma}{4} \int_{t_i}^{t_f} dt \frac{\dot{\omega}^2}{\omega} - \frac{1}{2}T[\ln \omega]_{t_i}^{t_f} + \frac{1}{2}[\lambda\omega]_{t_i}^{t_f} \tag{2.10}$$

$$= W_{irr} - T\Delta S + \Delta E. \tag{2.11}$$

As indicated, we can identify three different contributions to the work performed on the system. The first term is the irreversible work lost during the process, the second part is the energy stored in the entropy change of the system and the third part amounts to the system's total energy change. The irreversible work is now to be minimized; therefore the integral in (2.11) is treated as an effective action for which Lagrangian equations of motion can be derived

$$L_{\text{eff}} = \frac{\gamma}{4} \frac{\dot{\omega}^2}{\omega}, \quad \frac{\partial L_{\text{eff}}}{\partial \omega} = \frac{d}{dt} \frac{\partial L_{\text{eff}}}{\partial \dot{\omega}} \quad \Rightarrow \quad \dot{\omega}^2 - 2\ddot{\omega}\omega = 0. \tag{2.12}$$

To solve these equations, boundary conditions have to be imposed on the variances. These are chosen as

$$\omega(0) = \omega_a, \quad \omega(t_1) = \omega_b, \quad \omega(t_1 + t_3) = \omega_a, \tag{2.13}$$

which means that during the coupling to the hot bath the variance is driven from ω_a to ω_b and during the coupling to the cold bath the variance is driven from ω_b to ω_a. Thereby a continuous periodic change of the distribution is ensured. Note that the boundary conditions have been imposed on the distribution of particles rather than on the control parameter λ, which might seem to be the more natural choice in the light that λ is the parameter that can be experimentally controlled. However, optimal protocols derived by imposing conditions on λ lead to a process in which the probability distribution is not changed at all. The irreversible work is thereby minimized, but in a cyclic process to which we want to apply the protocols this produces zero work output. This of course is to be avoided.

With the boundary conditions (2.13) the Lagrange equation (2.12) can be solved, and the resulting optimal motion for the variances at the hot and cold bath respectively read

$$
\begin{aligned}
\omega_1^*(t) &= \omega_a(1 + (\sqrt{\omega_b/\omega_a} - 1)t/t_1)^2 & \text{for} \quad 0 < t < t_1, \\
\omega_3^*(t) &= \omega_b(1 + (\sqrt{\omega_a/\omega_b} - 1)(t - t_1)/t_3)^2 & \text{for} \quad t_1 < t < t_1 + t_3.
\end{aligned}
\tag{2.14}
$$

Equation (2.14) can be translated back with equation (2.9) to give the optimal protocols for the control parameter λ

$$
\begin{aligned}
\lambda_1(t)^* &= \frac{t_1^2 T_h + t_1(\omega_a - \sqrt{\omega_a \omega_b})\gamma - t(\sqrt{\omega_b} - \sqrt{\omega_a})^2\gamma}{[t_1\sqrt{\omega_a} + t(\sqrt{\omega_b} - \sqrt{\omega_a})]^2} & \text{for} \quad 0 < t < t_1, \\
\lambda_3(t)^* &= \frac{t_3^2 T_c + t_3(\omega_b - \sqrt{\omega_a \omega_b})\gamma - (t - t1)(\sqrt{\omega_a} - \sqrt{\omega_b})^2\gamma}{[t_3\sqrt{\omega_b} + (t - t_1)(\sqrt{\omega_a} - \sqrt{\omega_b})]^2} & \text{for} \quad t_1 < t < t_1 + t_3.
\end{aligned}
\tag{2.15}
$$

One such protocol for a typical parameter set is depicted in figure 2.2(b). Using the solution (2.14) in the expression for the work (2.11) and including the additional contribution to the work due to the sudden jumps of the potential at the bath change in equation (2.1), one can write down an expression for the total work output of the engine:

$$
-W = \frac{1}{2}(T_h - T_c)\ln\frac{w_b}{w_a} - \gamma\left(\frac{1}{t_1} + \frac{1}{t_3}\right)(\sqrt{w_b} - \sqrt{w_a})^2.
\tag{2.16}
$$

The heat uptake during the coupling to the hot bath at temperature T_h is

$$
Q^{(1)} = T_h \Delta S - W_{irr}^{(1)} = \frac{1}{2}T_h \ln\frac{w_b}{w_a} - \frac{\gamma}{t_1}(\sqrt{w_b} - \sqrt{w_a})^2.
\tag{2.17}
$$

Combining equations (2.17) and (2.16) allows us to calculate the efficiency of the heat engine

$$
\eta = \frac{-W}{Q^{(1)}} = \frac{\frac{1}{2}(T_h - T_c)\ln\frac{w_b}{w_a} - \gamma\left(\frac{1}{t_1} + \frac{1}{t_3}\right)(\sqrt{w_b} - \sqrt{w_a})^2}{\frac{1}{2}T_h \ln\frac{w_b}{w_a} - \frac{\gamma}{t_1}(\sqrt{w_b} - \sqrt{w_a})^2}.
\tag{2.18}
$$

Taking the limit of infinite coupling times, which amounts to a quasi-static process, we recover the Carnot efficiency

$$
\lim_{t_1, t_3 \to \infty} \eta = 1 - \frac{T_c}{T_h}.
\tag{2.19}
$$

As pointed out in section 1.3, infinite coupling times lead to a zero power output, but we actually want to maximize the power output for the heat engine. We therefore treat the coupling times t_1 and t_3 as free parameters and maximize the power $P = -W/(t_1 + t_3)$

with respect to them. The conditions $\partial P/\partial t_1 = 0$ and $\partial P/\partial t_3 = 0$ then lead to the optimal coupling times

$$t_1^* = t_3^* = \frac{8\gamma(\sqrt{w_b} - \sqrt{w_a})^2}{(T_h - T_c)\ln\frac{w_b}{w_a}}.$$ (2.20)

Using t_1^* and t_3^* in the expression for the efficiency (2.18) we finally arrive at the important efficiency at maximum power

$$\eta^* = \frac{2(T_h - T_c)}{3T_h + T_c} = \frac{\eta_C}{2 - \eta_C/2}.$$ (2.21)

Figure 2.2(a) shows η^* over the ratio of bath temperatures T_c/T_h and compares it to the Curzon-Ahlborn efficiency (1.71) derived for a classical heat engine at maximum power. For small temperature differences η^* reduces to the Curzon-Ahlborn efficiency, but it deviates for large temperature differences.

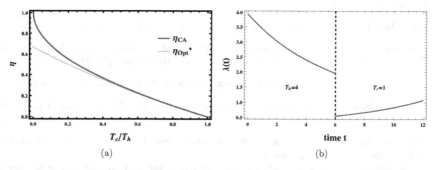

Figure 2.2: (a) Curzon-Ahlborn efficiency η_{CA} and efficiency at maximum power η_{Opt}^* for a stochastic heat engine with optimally driven harmonic potential [22]. For small temperature differences both coincide while for large temperature differences they differ significantly. (b) Typical optimal protocol for a parameter set $\gamma = 1, w_a = 1, w_b = 2, \Delta t = 12, T_h = 4, T_c = 1$.

2.3 Breakdown of the Overdamped Limit

2.3.1 Motivation

In most processes observable in nature friction is so high that the overdamped assumption (see section 1.1.5) is justified. Also the experimentally realized stochastic heat engine by Blickle and Bechinger [23] described in chapter 3.1 operated in this regime due to the embedding in water which led to a high damping constant. However, recent advances in optical trapping techniques now allow to access the underdamped regime where inertia effects become important. We especially have in mind the optical cavity trap [36] described in chapter 3.2. The cavity is filled with a dilute gas. Naturally, low pressure leads to low damping. To describe a heat engine realized by such a trap accurately one has to generalize the theory outlined in section 2.2 to the underdamped regime.

2.3.2 Numerical Study

In a first step we performed a numerical study in which we simulated the stochastic heat engine presented in section 2.2. Thereby, we compared the full dynamics including the inertia term and the overdamped dynamics, where the inertia term was neglected. We used the method of Langevin simulations, whereby the Gaussian white noise was created via a Box-Muller algorithm [48] and averages were taken over at least $3 \cdot 10^5$ trajectories. Thermodynamic quantities were calculated with the help of stochastic energetics (see section 1.2). Details on the simulation technique can be found in appendix A.3.

When considering the full dynamics, we essentially have four important timescales that determine the system.

1. The relaxation timescale of the velocity characterized by $\tau_{rel} \sim 1/\gamma$.

2. The timescale τ_λ on which we change the control parameter $\lambda(t)$ of the system. It may be characterized by the total cycle time of the engine $\tau_\lambda \sim \Delta t$.

3. The oscillation period of the Brownian particle in the harmonic trap $\tau_{osc} \sim 1/\sqrt{\lambda}$.

4. The timescale τ_T on which the temperature is changed. In the model at hand the switching is assumed to be instantaneous, i.e $\tau_T = 0$ [1].

The overdamped limit is valid when the relaxation time of velocity, τ_{rel}, is much shorter than all other characteristic timescales.

We implemented the optimal protocols derived in the overdamped regime (equation (2.15)) and compared the predictions for power, efficiency and variances to the numerical results of the full dynamics, including inertia. All simulation results are presented in dimensionless units and can be rescaled to apply to a specific experimental setup with equation (A.2). We fixed the temperature difference between hot and cold bath by a factor of four, i.e. $T_h = 4, T_c = 1$ and chose the boundary conditions for the variances as $\omega_a = 1, \omega_b = 2$.

Case $\gamma = 10$

In figure 2.3(a) we see the power as a function of the cycle time Δt. For the depicted range of Δt, the relaxation timescale for the velocity is much smaller than the other timescales, i. e. $\tau_{rel} \ll \tau_\lambda$ and $\tau_{rel} \ll \tau_{osc}$, and the overdamped assumption is fairly justified. As a result the curves for the over-, and underdamped cases are close together. In particular the prediction at maximum power, where we ideally would like to run the engine, is applicable. Figure 2.3(b) shows the efficiencies with respect to the cycle duration. We observe that for long

---

[1]Since the temperature is changed instantaneously, at the bath change the overdamped limit always fails.

cycle times the efficiency for the underdamped case does not approach the predicted Carnot efficiency η_C but a lower efficiency that we denote by η_{C-Leak}. This can be explained as follows. To reach Carnot efficiency the engine has to be run reversibly, which corresponds to a quasi-static change for the probability distribution of the Brownian particle. The bath change is performed instantaneously ($\tau_T = 0$), i.e. faster than the relaxation time of the velocity τ_{rel}. This irreversible relaxation leads to an extra contribution Q_{Leak} in the heat exchange which is not taken into account in the overdamped limit. Here, the heat transfer between bath and system only depends on the distribution in position space and the velocities after the bath switching are instantly assumed to be Maxwell distributed around the thermal velocity $v_{th} = \sqrt{T}$. For long enough cycle times this heat leakage is the energy difference in average kinetic energy

$$Q_{Leak} = \frac{1}{2}k_B(T_h - T_c),\qquad(2.22)$$

which leads to the corrected Carnot-Leak efficiency

$$\eta_{C-Leak} = \frac{-W}{Q^{(1)} + Q_{Leak}}.\qquad(2.23)$$

To recover Carnot efficiency in the underdamped regime, one would have to make the bath change truly adiabatic by changing the bath temperature infinitely slowly, thereby rendering the process reversible [49].

In figure 2.4(b) the mean square velocity is depicted for a single cycle with duration $\Delta t = 12$, which corresponds to the maximum in the power curve. Because of the relatively high damping, the velocity relaxation, whose timescale is determined by $1/\gamma$, is very fast compared to the cycle duration. Figure 2.4(a) shows the mean square displacement which for both regimes almost coincides.

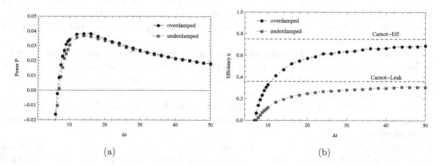

(a) (b)

Figure 2.3: (a) Power and (b) efficiency over duration of cycle time for a stochastic heat engine with implemented optimal protocols for the overdamped case. Parameter set: $\gamma = 10$, $T_h = 4$, $T_c = 1$, $\omega_a = 1$, $\omega_b = 2$.

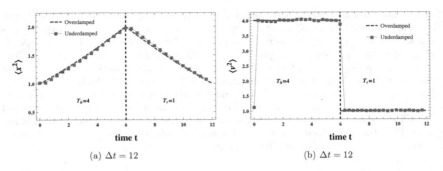

(a) $\Delta t = 12$ (b) $\Delta t = 12$

Figure 2.4: (a) Mean square displacement and (b) mean square velocity over time for a fixed cycle duration of Δt. Parameter set: $\gamma = 10$, $T_h = 4$, $T_c = 1$, $\omega_a = 1$, $\omega_b = 2$.

Case $\gamma = 1$

By decreasing the damping we shift the ratio of timescales in the system.

In figure 2.5(a) we see again the power plotted over the duration of the cycle time, but now the difference between overdamped and underdamped simulation is more pronounced. Asymptotically for large cycle times, the underdamped result approaches the overdamped result, because at some point the overdamped assumption $\tau_{rel} \ll \tau_\lambda$ becomes true. But in the most important range of cycling times Δt near to where (according to the overdamped limit) the power maximum lies, overdamped and underdamped results differ significantly. Considering the full dynamics, the power output in this range even becomes negative, meaning that the engine consumes work instead of performing. For even shorter cycle times, when the power output in the overdamped regime becomes negative, the power for the underdamped case goes to zero. Here the cycle times are so short ($\tau_\lambda < \tau_{rel}$) that inertia effects prevent the particle distribution to react to neither the potential nor the temperature change.

In figure 2.5(b) we see that the efficiency for the underdamped case becomes positive at longer cycle times, which is due to the fact that also the power output becomes positive at a later time. Again the efficiencies saturate at the Carnot and the Carnot-leak efficiency respectively for over-, and underdamped case.

In order to illustrate the three regimes of cycle times described above, i.e. $\tau_{rel} > \tau_\lambda$, $\tau_{rel} \sim \tau_\lambda$ and $\tau_{rel} \ll \tau_\lambda$, we look at the mean square velocity and mean square displacement plots in figure 2.6.

Case $\Delta t = 0.5$, $(\tau_{rel} > \tau_\lambda)$, Fig. 2.6(a) and 2.6(b): We see that before the velocity can relax to the bath temperature the coupling to the bath is switched. Due to the inertial effects the mean square displacement hardly changes during the cycle which leads to zero work and power output.

Case $\Delta t = 1.5$, $(\tau_{rel} \sim \tau_\lambda)$, Fig. 2.6(c) and 2.6(d): In the overdamped case this cycle time corresponds to maximum power output, whereas in the underdamped case the power output becomes negative. If we look at the underdamped mean square velocity, we see that just when it has relaxed to the thermal velocity, the temperature bath is changed. During the relaxation process, mean square velocity and position are correlated and the relaxation of velocity translates into a counter-directed motion of the mean square position. The protocol $\lambda(t)$ is therefore not using the particle motion to extract work, but on the opposite has to perform against it. This is the source of the negative power output.

Case $\Delta t = 50$, $(\tau_{rel} \ll \tau_\lambda)$, Fig. 2.6(e) and 2.6(f): Finally, at a very long cycle time the predictions for over and underdamped limit agree quite well again. Only directly after the temperature bath change, the relaxation of the mean square velocity translates into a large peak in the mean square position that quickly decays. During this relaxation process oscillations start to occur, whose origin is the fact that $\tau_{rel} > \tau_{osc}$. For the shorter cycle times of $\Delta t = 0.5$ and $\Delta t = 1.5$ these were not visible, because the cycle time itself was too short, i.e. $\Delta t < \tau_{osc}$.

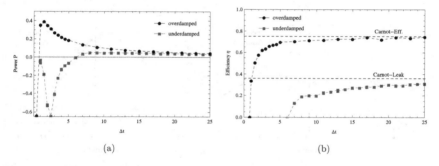

(a) (b)

Figure 2.5: (a) Power over duration of cycle time. (b) Efficiency over duration of cycle time. Parameter set: $\gamma = 1$, $T_h = 4$, $T_c = 1$, $\omega_a = 1$, $\omega_b = 2$.

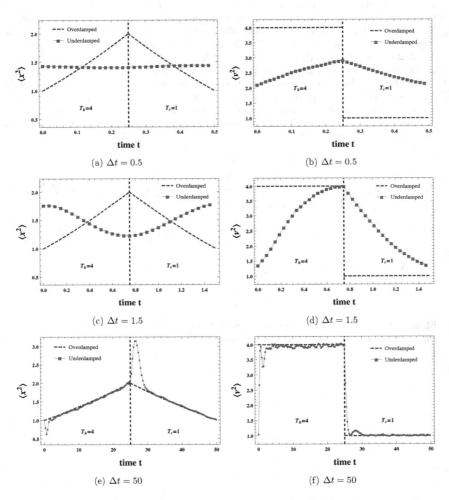

Figure 2.6: (a), (c), (e) Mean square displacement and (b), (d), (f) mean square velocity over time for fixed cycle durations of $\Delta t = 0.5, 1.5$ and 50 respectively. Parameter set: $\gamma = 1$, $T_h = 4$, $T_c = 1$, $\omega_a = 1$, $\omega_b = 2$.

Case $\gamma = 0.1$

We now delve even further into the underdamped limit. In figure 2.7(a) and 2.7(b) we see the mean square displacement and mean square velocity plots for a very long cycle time of $\Delta t = 500$. Two interesting features stick out. First we observe that the mean square velocity exceeds the thermal velocity even after relaxation. Secondly we observe

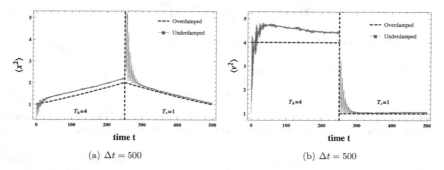

(a) $\Delta t = 500$ (b) $\Delta t = 500$

Figure 2.7: (a) Mean square displacement and (b) mean square velocity over time for a fixed cycle duration of $\Delta t = 500$. Parameter set: $\gamma = 0.1$, $T_h = 4$, $T_c = 1$, $\omega_a = 1$, $\omega_b = 2$.

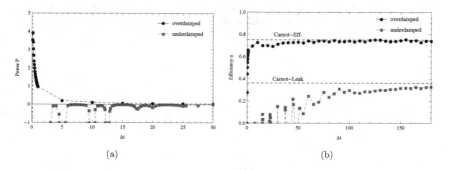

(a)   (b)

Figure 2.8: (a) Power and (b) efficiency over duration of cycle time for a stochastic heat engine with implemented optimal protocols for the overdamped case. Parameter set: $\gamma = 0.1$, $T_h = 4$, $T_c = 1$, $\omega_a = 1$, $\omega_b = 2$.

pronounced oscillations after the bath changes which come about due to the fact that $\tau_{rel} > \tau_{osc}$. These oscillations translate even into the power and efficiency plots in figures 2.8(a) and 2.8(b). It therefore seems necessary to carefully adjust engines operating at cycle times comparable to the relaxation timescale of the velocity $\sim 1/\gamma$ to the phase of these oscillations to obtain optimal results for power output and efficiency.

The origin of the oscillations can at least qualitatively be explained by an analytical argument. If we assume the harmonic potential to be static at the bath change, analytical expressions for the transition probability and therefore for the probability distribution can be found [25]. We here just outline the procedure. The drift and diffusion matrix for a constant control parameter λ_p are given by

$$\mathbf{D}^{(1)} = \begin{pmatrix} 0 & -1 \\ \lambda_p & \gamma \end{pmatrix}, \quad \mathbf{D}^{(2)} = \begin{pmatrix} 0 & 0 \\ 0 & T\gamma \end{pmatrix}. \tag{2.24}$$

The transition probability then has the shape of a Gaussian

$$P(x, v, t | x', v', 0) = \frac{1}{2\pi\sqrt{\det\sigma}} \exp\left\{ \frac{1}{2\det\sigma} \left[-\sigma_{vv}(x - x')^2 \right. \right.$$
$$\left. \left. +2\sigma_{xv}(x - x')(v - v') - \sigma_{xx}(v - v')^2 \right] \right\}. \tag{2.25}$$

After a spectral decomposition of the drift matrix into a biorthogonal basis[25] the variances of this Gaussian can be shown to depend on eigenvalues of the drift matrix

$$\sigma_{ij} = \sigma_{ij}(\lambda_1, \lambda_2), \quad \lambda_{1,2} = \frac{1}{2}\left(\gamma \pm \sqrt{\gamma^2 - 4\lambda_p}\right). \tag{2.26}$$

For γ smaller than a critical γ_c

$$\gamma_c = 2\sqrt{\lambda_p m} \tag{2.27}$$

these eigenvalues become complex, which results in oscillations of the relaxing distribution. Since our control parameter $\lambda(t)$ is time dependent, this result is not exact but it explains the order of magnitude of γ_c at which oscillations occur very well.

3
Experimental Setups in Overdamped and Underdamped Regimes

3.1 Realization of a Stirling Engine in a Colloidal System

In section 2.1 we introduced a model for a stochastic heat engine operating on the micro scale. Such a heat machine has indeed been experimentally realized by Blickle and Bechinger [23] in 2012. They thereby proved that micromechanical machines that convert thermal to mechanical energy are not merely a theoretical playground, but provide a feasible design for future applications. We therefore review their experiment in some detail, a sketch of the working principle is given in figure 3.1. The experiment consisted in the single particle manipulation of a melamine bead with a diameter of 2.94 μm which was, because of its small size, subject to thermal fluctuations. The theoretical description of such a particle has to be performed in the framework of stochastic dynamics introduced in 1.1. The bead was trapped by a highly focused infrared laser beam[1]. The finite polarizability of the particle leads to an interaction with the laser field, that results in an optical gradient force which depends on the field intensity. The particle is therefore trapped around the intensity maximum which amounts to a parabolic potential $V(\lambda(t), x) = 1/2\lambda(t)x^2$ exerted on the particle, where x is the radial distance from the trapping center and the value of $\lambda(t)$ could be controlled by the laser intensity. As a heat bath served a vitreous sample cell filled with water. For a spherical particle the damping constant is proportional to the viscosity of the thermal environment, which for water is very large. The experiment was therefore performed well in the overdamped limit. For a heat engine to work, one needs two bathes at different temperatures. On small length scales, it is experimentally very difficult to achieve thermal isolation between different heat reservoirs. Rather than coupling the bead periodically to a hot and cold environment, the temperature of the surrounding liquid was suddenly changed by a second laser beam whose wavelength was matched to an absorption peak of water at a wavelength of 1.455nm. The spot size was thereby adjusted as to avoid temperature gradients in the moving region of the particle and the beam was only applied to a thin layer in which the particle moved, so that the thermal relaxation time to the cold temperature was negligible as soon as the laser was switched of. Hence, this corresponds to the instantaneous switch of the temperature bath in the model of section

---

[1]also called optical tweezer, for details on how such a trap works see the book on optical trapping techniques by Ashkin [47].

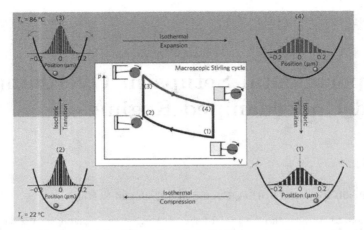

Figure 3.1: Working principle of the microscopic Stirling engine that was experimentally realized by Blickle and Bechinger. The harmonic trapping potential was opened and closed by a linear driving protocol during the isothermal phases. The isochoric transitions were achieved by instantaneously heating or cooling the temperature bath. Image taken from [23].

2.1. Thermodynamic quantities such as heat and work were calculated with the formulas of stochastic energetics (see section 1.2) for which only the particle trajectories had to be measured. These were imaged on the chip of a charge-coupled device camera using a microscope with a ×50 objective. As such, the experimental setup closely corresponded to the theoretical model described in section 2.1, with the difference that the protocols were not optimized to reduce the work lost to irreversible entropy production. Instead, merely a linear protocol was investigated $\dot{\lambda} = const.$, which was run from λ_{min} to λ_{max} during the cold phase, and reversely from λ_{max} to λ_{min} during the hot phase. The coupling times to hot and cold bath were chosen equally long and the total cycle time varied between 2 and 50 seconds. In figure 3.2 the measured results for the work and power output, as well as the obtained efficiencies are depicted for different total cycle times. As predicted in the theoretical model, the power output exhibits a maximum at some optimal cycle time. This comes about due to the trade-off between large cycle times where dissipation is negligible and the work output is large, and high dissipation at small cycle times where the particle is driven far away from equilibrium caused by the rapid change of the trapping potential. The efficiency curve relaxes to some finite efficiency for large cycle times as we have also seen in our numerical study in chapter 2.3.2. At maximum power the efficiency corresponds within the measurement errors well to the Curzon-Ahlborn efficiency.

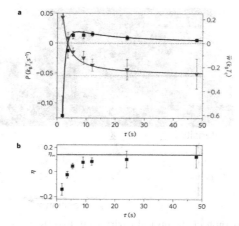

Figure 3.2: Experimental results by Blickle and Bechinger for a microscopic heat engine with experimental conditions $T_h = 76°C, T_c = 22°C, \lambda_{max}/\lambda_{min} = 4.4$. **a**: Averaged work $\bar{w}$ and power P as a function over cycle time τ. **b**: The efficiency η of the heat machine versus cycle time. The solid lines are (aymptotic) fits to the measured data. Image taken from [23].

3.2 Optical Cavity Trap

During the work on this thesis, we were in close communication to the experimental group of Nikolai Kiesel and Markus Aspelmeyer at the University of Vienna. In their laboratory they achieved to set up a very sophisticated optical cavity trap in which single nanoparticles can be confined and manipulated [36]. They used silica nanospheres with a diameter of around 260 nm. An important feature of their experiment is not only the ability to trap particles of such a small size, but also to be able to impose an additional optical damping that manipulates the particle's kinetic energy. A picture of a nanoparticle trapped in the optical cavity can be seen in figure 3.3a. Similar trap setups employing the same experimental technique were achieved in [37, 38, 39].

In the following we will briefly outline how the trapping mechanism works; for a more detailed description comprising the derivation of formulas for all optomechanical interactions we refer to the original paper on the experiment by the group [36]. The spatial mode of an optical cavity leads to a standing-wave intensity distribution along the cavity axis. Due to its finite polarizability ζ, a mounted particle is trapped at one of the maxima x_0 of the standing wave, similar to what happens in an optical tweezer. In addition the cavity resonance frequency is shifted by an amount $U_0(x_0)$ due to Rayleigh scattering off the particle and into the cavity mode. The resulting interaction Hamiltonian reads

$$H_{\text{int}} = -\hbar U_0(x_0)\hat{n}\sin^2(kx_0 + k\bar{x} + k\hat{x}), \tag{3.1}$$

where $\hat{x}$ is the position operator of the trapped particle, k is the wave number of the cavity

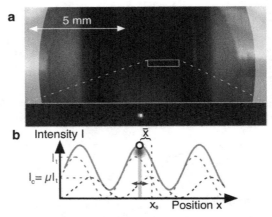

Figure 3.3: **a**: Photograph of the optical cavity trap used by Kiesel *et. al.* The trapped particle has a size of 300 *nm* and is confined by the superposition of two standing laser waves. **b**: Scheme of the trapping technique via superposition of standing waves. The effective harmonic potential is controlled mainly by the trapping beam. The phase shift $\bar{x}$ between trapping and control beam allows for the introduction of an additional optical damping. Pictures taken from Ref. [36].

mode, $\hat{n}$ is the cavity photon number operator and $\bar{x}$ is the mean displacement of the particle from the maximum at x_0. For the case of a single optical cavity mode $\bar{x} = 0$ and for small displacements, only coupling terms quadratic in $\hat{x}$ are relevant, which essentially means that the particle moves in a harmonic potential. In the actual experiment two slightly different cavity modes created by two different lasers, referred to as trapping beam and control beam, resonate in the optical cavity and form standing waves (see the sketch in figure 3.3b). The mounted nanoparticle is then trapped at one of the maxima of the combined intensity distribution. The trapping beam has a much higher intensity than the control beam so that the combined maximum at x_0, around which the confined particle oscillates, is mainly determined by the former. The particle couples to the control field at a shifted position $\bar{x} \neq 0$ (see figure 3.3b), which in the Lamb-Dicke regime ($k^2 \langle \hat{x}^2 \rangle \ll 1$) yields a strong linear optomechanical coupling [50]. Detuning the control field from the cavity resonance by a frequency $\Delta = \omega_{\mathrm{cav}} - \omega_{\mathrm{contr}}$ can then be used to introduce an additional optomechanical damping.

The trajectory of the trapped nanoparticle can theoretically be described as Brownian motion in a potential of cylindrical symmetry

$$V(x, y, z) = -V_0 e^{-\frac{r^2}{r_c^2}} \cos^2(kz) \quad \text{with} \quad r^2 = x^2 + y^2, \tag{3.2}$$

where x_c, V_0 and k are parameters that can be controlled via tuning of wavelength, intensity and mode shape of the trapping and the control laser beams. The exponential part in the potential originates in the Gaussian profiles of the lasers and the cosine part comes from the intensity distribution of the standing wave along the cavity axis. The resulting Langevin equations read

$$dv_x = \left[-\gamma_{th}v_x - \frac{2V_0}{r_c^2}xe^{-\frac{r^2}{r_c^2}}\right]dt + \sqrt{\frac{2\gamma k_B T}{m}}dW_x$$

$$dv_y = \left[-\gamma_{th}v_y - \frac{2V_0}{r_c^2}ye^{-\frac{r^2}{r_c^2}}\right]dt + \sqrt{\frac{2\gamma k_B T}{m}}dW_y \qquad (3.3)$$

$$dv_z = \left[-\left(\gamma_{th} + \gamma_{opt}e^{-\frac{r^2}{r_c^2}}\right)v_z - 2kV_0\sin(kz)\cos(kz)e^{-\frac{r^2}{r_c^2}}\right]dt + \sqrt{\frac{2\gamma k_B T}{m}}dW_z.$$

Note that in z-direction along the cavity axis an additional optical damping γ_{opt} has been introduced that can be manipulated with the control beam and therefore also couples to the Gaussian intensity profile.

Currently the experiment operates at room temperature of $T = 295K$. In this regime the particle motion is very small compared to the length scales of the potential ($r^2/x_c^2 \ll 1$, $kz \ll 1$) so that an harmonic approximation of the potential is sufficient

$$V(x,y,z) \approx -V_0 + \frac{1}{2}m\omega_r^2(x^2+y^2) + \frac{1}{2}m\omega_z^2 z^2 \quad \text{with} \quad \omega_r = \sqrt{\frac{2V_0}{mr_c^2}}, \quad \omega_z = \sqrt{\frac{2V_0 k^2}{m}}. \quad (3.4)$$

Typical experimental values for the parameters are[2]

$$m = 10^{-17}kg, \quad r_c = 40 \cdot 10^{-6}m, \quad k = 2\pi \cdot 10^6\frac{1}{m}, \quad V_0 = 2 \cdot 10^{-19}J, \qquad (3.5)$$

which gives trapping frequencies of

$$\frac{\omega_r}{2\pi} \approx 800Hz, \quad \frac{\omega_z}{2\pi} \approx 200kHz. \qquad (3.6)$$

Due to the large difference in trapping frequencies, the system may in a theoretical description (if convenient) be reduced to a one-dimensional system.

In contrast to the experiment by Blickle and Bechinger [23] which was performed in water, the optical cavity is filled with a dilute gas and the experiment is therefore designed to run in the underdamped regime. This leads to a much smaller damping constant for the particle ($\gamma_{th}/m = 5kHz$). Comparing this to the typical trapping frequency ω_z we see that $\tau_{rel} > \tau_{osc}$. A stochastic heat engine implemented in this trap setup would therefore not operate in the overdamped limit. An extension of the theory of optimal protocols discussed in section 2.2 in the overdamped limit is hence necessary.

---

[2]The source of information regarding the typical experimental parameters, possible parameter ranges and which parameters can be controlled in the experiment, is direct communication with Nikolai Kiesel who is running the experiment in Vienna.

3.2.1 Effective Heat Engine via Optical Damping

In the harmonic approximation, the motion of a particle along the optical cavity axis may be described by a one-dimensional Langevin equation, which in rescaled units reads

$$\ddot{x} = -(\gamma_{\text{th}} + \gamma_{\text{opt}})\dot{x} - \lambda x + \sqrt{2\gamma_{\text{th}}T}\xi(t). \tag{3.7}$$

To realize a stochastic heat engine two different temperature baths are necessary, but in the optical cavity trap it is experimentally very challenging to change the temperature of the surrounding gas quickly. It is, however, possible to manipulate the optical damping. We therefore rewrite equation (3.7) in the form of the familiar Langevin equation without optical damping by introducing an effective total damping which automatically leads to an effective temperature

$$\ddot{x} = -\gamma_{\text{eff}}\dot{x} - \lambda x + \sqrt{2\gamma_{\text{eff}}T_{\text{eff}}}\xi(t) \tag{3.8}$$

$$\text{with} \quad \gamma_{\text{opt}} = f\gamma_{\text{th}}, \quad \gamma_{\text{eff}} = \gamma_{\text{th}}(1+f), \quad T_{\text{eff}} = \frac{T}{1+f}. \tag{3.9}$$

It is now possible to change the effective temperature by changing the optical damping, which is much more convenient from an experimentalist's point of view. A heat engine could then be realized by switching the optical damping on and off, resulting in

$$\begin{aligned} \text{cold bath:} \quad & \gamma_{\text{opt}} = f\gamma_{\text{th}}, \quad \rightarrow \quad T_c = \frac{T}{1+f}, \quad \mu_c = \frac{1}{\gamma_{\text{th}}(1+f)}, \\ \text{hot bath:} \quad & \gamma_{\text{opt}} = 0, \quad \rightarrow \quad T_h = T, \quad \mu_h = \frac{1}{\gamma_{\text{th}}}, \end{aligned} \tag{3.10}$$

where we introduced the mobility $\mu = 1/\gamma$. Note that this effectively corresponds to switching between two temperature baths with different dampings.

In the overdamped limit the optimal protocol for the control parameter λ that minimizes the irreversible work has been derived by Seifert *et al.* [22] and was presented in equation (2.15). We performed a similar calculation but adjusted for the case of two different dampings. With equation (3.10) the overdamped optimal protocols become

$$\lambda_h(\tau) = \frac{t_1((1+f)^2 t_1 T_c \mu_c + \omega_a - \sqrt{\omega_a\omega_b}) - \tau(\sqrt{\omega_b} - \sqrt{\omega_a})^2}{(1+f)\mu_c(t_1\sqrt{\omega_a} + \tau(\sqrt{\omega_b} - \sqrt{\omega_a}))^2}, \tag{3.11}$$

$$\lambda_c(\tau) = \frac{t_3^2 T_c u_c + t_3(\omega_b - \sqrt{\omega_b\omega_a}) - (\tau - t_1)(\sqrt{\omega_b} - \sqrt{\omega_a})^2}{\mu_c(t_3\sqrt{\omega_b} + (\tau - t_1)(\sqrt{\omega_b} - \sqrt{\omega_a}))^2}, \tag{3.12}$$

where ω_a and ω_b are the boundary conditions imposed on the mean square position. The total work output during a complete cycle of the engine results in

$$-W_{tot} = \frac{[t_1(1+f) + t_3](\sqrt{\omega_b} - \sqrt{\omega_a})^2}{(1+f)t_1 t_3 \mu_c} - \frac{1}{2}fT_c\ln(\omega_b/\omega_a), \tag{3.13}$$

and the power is respectively $P = W_{tot}/(t_1 + t_3)$.

Unlike in the case of real temperature baths with the same damping, the optimal coupling times to hot and cold bath that maximize power turn out to not be equal

$$t_1^* = \frac{4(1 + \sqrt{1+f})(\sqrt{\omega_b} - \sqrt{\omega_a})^2}{f(1+f)T_c\mu_c\ln(\omega_b/\omega_a)}, \quad t_3^* = \sqrt{1+f}t_1^*. \tag{3.14}$$

The heat uptake at the hot bath is

$$Q^{(1)} = -\frac{1}{2}(1+f)T_c\ln\left(\frac{\omega_b}{\omega_a}\right) + \frac{(\sqrt{\omega_b} - \sqrt{\omega_a})^2}{t_1(1+f)u_c} - Q_{leak}, \tag{3.15}$$

where Q_{leak} is the contribution due to the instantaneous relaxation of the velocity as explained in section 2.3.2 and is given by $Q_{leak} = fT_c/2$. Inserting the coupling times for maximum power in the formula for the efficiency $\eta = -W_{tot}/Q^{(1)}$, we arrive at the efficiency at maximum power

$$\eta^* = \frac{f\ln(\omega_b/\omega_a)}{(3 + 2f - \sqrt{1+f})\ln(\omega_b/\omega_a) + 2f}, \tag{3.16}$$

which yields a maximum for an optimal ratio between optical and thermal damping

$$f^* = 4(4 + 3\sqrt{2}) \approx 32.97. \tag{3.17}$$

Note that the factor f can be expressed as $f = T_h/T_c - 1$ and determines the ratio of effective temperatures. It is therefore quite surprising that the efficiency at maximum power exhibits a maximum at an optimal value for f. It comes about due to the fact, that f does not only change the temperature difference but also the difference in the damping constant between the two baths. This result yields a qualitative difference from the efficiency at maximum power of a heat engine operated with two real temperature baths, where η^* is a monotonically increasing function of the ratio of temperatures T_h/T_c (see equation (2.21)).

3.2.2 Numerical Study

The experimental group around Nikolai Kiesel and Markus Aspelmeyer has expressed a strong interest in realizing a heat engine via optical damping. Since they have very good control over the trapping lasers, they in particular want to investigate optimal protocols that maximize the power output. Therefore we numerically studied the model described in the last section including the full dynamics and carefully paid attention to simulate experimentally accessible parameters. The objective here was to determine whether or not their trap setup is theoretically still describable in the overdamped approximation for which we presented the theory in the last section. Currently, the experimentally feasible range of the ratio between optical and thermal damping $f = \gamma_{opt}/\gamma_{th}$ lies between 0 and 10. But there are also restrictions on the range of the control parameter $\lambda(t)$, i.e. $\lambda_{max}/\lambda_{min} \approx 4$. To be able to run the optimal protocols derived in the overdamped case (equation (3.12)) one has to adjust the boundary conditions ω_b and ω_a, the accessible range of optical damping

f and the range of the control parameter λ. For boundary conditions of $\omega_b/\omega_a = 1.5$ and an optical damping factor of $f = 1$ the optimal protocols are realizable for a wide range of cycle times in which we are interested.

In figure 3.4 the mean values for power and efficiency are presented. We see that only for very long cycle times the predictions derived in the overdamped limit are approached. Especially at optimal coupling times where the overdamped theory predicts maximum power and where we would finally like to run the engine, we are with the experimentally realistic parameters still far in the underdamped regime.

In the range of experimentally realistic parameters not only the crossing of timescales τ_{rel} and τ_λ is important but also the crossing between τ_{rel} and τ_{osc}. This can be seen in figure 3.5 in which the mean square displacement and mean square velocity are exemplarily depicted for a cycle time of $t = 1207[\sqrt{m_0/E_0}l_0]$ (in this regime $\tau_{\text{rel}} \sim \tau_\lambda$). We observe that after the bath change the variances relax towards the ones predicted in the overdamped limit. However, during the relaxation process fast oscillation occur that originate in the crossing of timescales $\tau_{\text{rel}} \gg \tau_{osc}$.

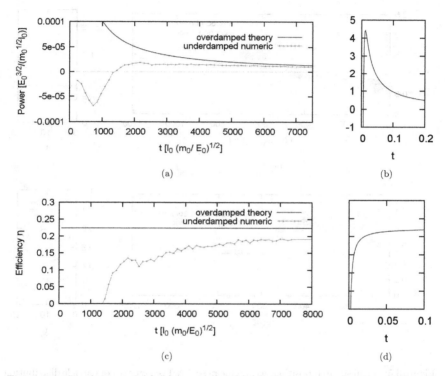

Figure 3.4: Simulations of (a),(b) power and (c),(d) efficiency as a function of time for a heat engine run by optical damping with experimentally feasible parameter conditions. Comparison of numerically solved full dynamics and overdamped analytical theory. Units are given in $E_0 = 300K \cdot k_B, l_0 = 10^{-8}m, m_0 = 10^{-17}kg$.

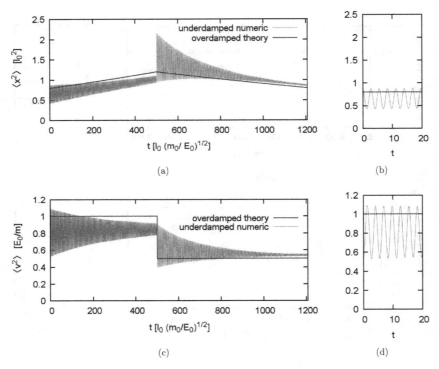

(a)   (b)

(c)   (d)

Figure 3.5: Mean squared position (a) and mean squared velocity (b) of particle distribution during one complete cycle in a heat engine run by optical damping with experimentally feasible parameters. Comparison of numerically solved full dynamics and overdamped analytical theory. Units are given in $E_0 = 300Kk_B, l_0 = 10^{-8}m, m_0 = 10^{-17}kg$. Oscillations occur due to the crossing of timescales $\tau_{\text{osc}} \ll \tau_{\text{rel}}$.

4

Stochastic Heat Engine - Underdamped Regime

New experimental techniques allow for the construction of microscopic heat engines that perform beyond the overdamped limit. In particular, the optical trap setup described in chapter 3.2 was designed such as to operate in the underdamped regime. In an accurate theoretical description the full dynamics has therefore to be taken into account. The probability distribution in phase space is consequently determined by the Klein-Kramers equation (1.22) which for a harmonic trapping potential (2.7) and in rescaled units reads

$$\frac{\partial \mathcal{P}(x,v,t)}{\partial t} = \left[-v\frac{\partial}{\partial x} + \frac{\partial}{\partial v}\left(\lambda(t)x + \gamma v\right) + \gamma T \frac{\partial^2}{\partial v^2} \right] \mathcal{P}(x,v,t). \tag{4.1}$$

Multiplying both sides with either x or v, integrating over the whole phase space and using integration by parts, the mean values vanish

$$\langle x \rangle = 0, \quad \langle v \rangle = 0. \tag{4.2}$$

Multiplying both sides of equation (4.1) with either x^2, v^2 or xv, integrating over the whole phase space, using integration by parts and equation (4.2), we can derive a set of coupled differential equations for the second moments:

$$\begin{aligned}
\dot{w}_x &= 2w_{xv} \\
\dot{w}_v &= -2\lambda w_{xv} - 2\gamma w_v + 2\gamma T \\
\dot{w}_{xv} &= w_v - \lambda w_x - \gamma w_{xv},
\end{aligned} \tag{4.3}$$

where we have defined

$$w_x = \langle x^2 \rangle, \quad w_v = \langle v^2 \rangle, \quad w_{xv} = \langle xv \rangle. \tag{4.4}$$

For Brownian motion in a harmonic potential it can be shown [25], that if the initial distribution is Gaussian the probability distribution remains Gaussian for all times, independently of the driving protocol $\lambda(t)$. It consequently is of the form

$$\mathcal{P}(x,v,t) = \frac{1}{2\pi\sqrt{w_x w_v - w_{xv}^2}} \exp\left[-\frac{w_v x^2 - 2w_{xv}xy + w_x v^2}{2(w_x w_v - w_{xv}^2)} \right], \tag{4.5}$$

where the variances w_x, w_v and w_{xv} are time-dependent and fulfill the set of coupled differential equations (4.3). Knowing the time-evolution of variances then completely determines the probability density distribution.

4.1   Optimization Problem

Along optimal protocols, the irreversible work lost due to the total entropy production is minimized. With equation (1.64) the expression for the irreversible work in the underdamped regime becomes

$$\langle W_{\text{irr}} \rangle = \langle W \rangle - \langle \Delta E \rangle + k_B T \langle \Delta S \rangle$$
$$= \int_{t_i}^{t_f} dt\, T \frac{dS}{dt} - \frac{d\langle Q \rangle}{dt} = \int_{t_i}^{t_f} dt \int \frac{\gamma}{\mathcal{P}} \left[v\mathcal{P} + T\frac{\partial \mathcal{P}}{\partial v} \right]^2 d\Gamma. \tag{4.6}$$

Substituting the Gaussian form of the probability distribution from equation (4.5) into equation (4.6) the irreversible work simplifies to

$$\langle W_{\text{irr}} \rangle = \gamma \int_{t_i}^{t_f} \left[w_v - 2T + \frac{T^2 w_x^2}{w_x w_v - w_{xv}^2} \right] dt. \tag{4.7}$$

Thinking in terms of a heat engine, we split the cycle into parts at constant temperature. Unlike in the overdamped case (see equation (2.11)) it is not possible to cast the irreversible work into a functional of one variable only. But with the help of equations (4.3) we can at least eliminate w_{xv} so that (4.7) becomes

$$W_{\text{irr}} = -2\gamma T(t_f - t_i) + \gamma \int_{t_i}^{t_f} \left[w_v + \frac{T^2 w_x^2}{w_x w_v - \dot{w}_x^2/4} \right] dt. \tag{4.8}$$

Furthermore, we use two of the equations in (4.3) to eliminate λ and w_{xv}, which gives us a constraint on the remaining variables w_x and w_y

$$\mathcal{G} = w_v \dot{w}_x - \frac{\gamma}{2}\dot{w}_x^2 - \frac{\dot{w}_x \ddot{w}_x}{2} + 2\gamma w_x (w_v - T) + w_x \dot{w}_v = 0. \tag{4.9}$$

The quantities w_v and w_x under the integral in equation (4.8) are not independent of each other, we therefore have to include the constraint $\mathcal{G}$ via a Lagrange multiplier α in the resulting Lagrangian

$$\mathcal{L} = \left[w_v + \frac{T^2 w_x^2}{w_x w_v - \dot{w}_x^2/4} \right] + \alpha(t)\mathcal{G}. \tag{4.10}$$

The optimization problem is then solved by solving the Euler Lagrange equations

$$\frac{\partial \mathcal{L}}{\partial w_x} + \frac{d^2}{dt^2} \frac{\partial \mathcal{L}}{\partial \ddot{w}_x} = \frac{d}{dt} \frac{\partial \mathcal{L}}{\partial \dot{w}_x}, \tag{4.11}$$

$$\frac{\partial \mathcal{L}}{\partial w_v} = \frac{d}{dt} \frac{\partial \mathcal{L}}{\partial \dot{w}_v}. \tag{4.12}$$

Unluckily the resulting set of nonlinear differential equations cannot be solved analytically. Furthermore also the numerical solution for specific parameter sets seems untraceable for the following reasons. The Lagrangian (4.10) has been derived for constant temperature, but the temperature in our model of the heat engine is a cyclic step function and we need to minimize the irreversible work during both isothermal processes. The natural way to tackle the problem is to impose endpoint boundary conditions at the beginning and the end of each isothermal process and patch the solutions at the bath changes to make the process periodic and smooth[1]. This does not constitute an initial value problem, but a boundary value problem, which generally is much more involved to solve numerically [51]. Furthermore, there is no guidance on how to choose the boundary conditions on the Lagrange multiplyer $\alpha(t)$ whose physical meaning is intransparent. After careful investigation, a numerical solution is out of scope of this work and we therefore focus on simplifying the equations by considering the limit of weak friction which corresponds to experimental conditions in the optical cavity discussed in section 3.2.

4.2 Low Friction Approximation

In the underdamped regime when friction is weak, the dynamics of the system is predominantly determined by the reversible part of the Fokker-Planck operator[2] and the energy of the system changes only slowly. In this limit a Fokker-Planck equation for the energy can be found, which for a static confining potential has been derived in [52]. We follow the same approach, but assume the potential to be time-dependent.

The energy of a particle and its total time derivative are generally given by

$$E = \dot{x}^2/2 + V(x,t),$$
$$\dot{E} = \dot{x}\ddot{x} + \frac{\partial V}{\partial x}\dot{x} + \frac{\partial V}{\partial t}. \tag{4.13}$$

Multiplying the usual Langevin equation (1.1) by $\dot{x}$:

$$\dot{x}d\dot{x} = -\left[\gamma\dot{x}^2 + \dot{x}\frac{\partial V(x,t)}{\partial x}\right]dt + \dot{x}\sqrt{2\gamma T}dW, \tag{4.14}$$

we can use equations (4.13) and (4.14) to derive a new set of Langevin equations in position and energy

$$dx = \pm\sqrt{2(E - V)}dt, \tag{4.15}$$
$$dE = \left[\frac{\partial V}{\partial t} - 2\gamma(E - V)\right]dt \pm 2\sqrt{(E - V)\gamma T}dW. \tag{4.16}$$

With equation (1.6) the drift and diffusion coefficients are given by

$$D_1^{(1)} = \pm\sqrt{2(E - V)}, \quad D_2^{(1)} = \frac{\partial V}{\partial t} - 2\gamma(E - V) + \gamma T,$$
$$D_{11}^{(2)} = 0, \quad D_{12}^{(2)} = D_{21}^{(2)} = 0, \quad D_{22}^{(2)} = 2\gamma T(E - V) \tag{4.17}$$

---

[1]This technique was followed in the overdamped case in section 2.2 where an analytical solution could be found.

[2]The reversible part of the FP-operator is just the Liouville streaming operator and corresponds to $L_0^{\dagger}$ in equation (A.4).

and the resulting Fokker-Planck equation becomes

$$\frac{\partial W}{\partial t} = \pm \frac{\partial}{\partial x} \sqrt{2(E-V)}W - \frac{\partial}{\partial E}\left[\frac{\partial V}{\partial t} - 2\gamma(E-V) + \gamma T\right]W + \frac{\partial^2}{\partial E^2}2\gamma T(E-V)W. \quad (4.18)$$

In a confining potential a trapped particle will oscillate from one side to the other. For small damping the energy is conserved during a large number of periods of these oscillations and the time the particle spends at a given point x is inversely proportional to the velocity $\dot{x} = \sqrt{2(E-V)}$ [52]. Therefore, for a fixed value of the energy, the conditional probability density may be approximated by

$$W(x,t|E) = \begin{cases} \text{const.}(E-V)^{-1/2}, & \text{for } V < E, \\ 0 & \text{for } V > E. \end{cases} \quad (4.19)$$

Note that if the potential is time-dependent, also the timescale τ_λ on which the potential is changed has to be slower than the timescale of oscillations τ_{osc} for this approximation to be valid. After normalizing, we can write the two-dimensional probability density as

$$W(x,E,t) = W(E)W(x,t|E) = \frac{W(E,t)}{2N(E,t)\sqrt{E-V}}, \quad N = \frac{1}{2}\int_{C(E)}\frac{dx}{\sqrt{E-V}}, \quad (4.20)$$

where the integration domain $C(E)$ is restricted to $V < E$. Inserting equation (4.20) into equation (4.18) and integrating over x, we obtain a one-dimensional Fokker-Planck equation depending solely on the energy. In the case of a harmonic potential $V = \lambda(t)x^2/2$ the integration domain for x runs from $-\sqrt{2E/\lambda}$ to $\sqrt{2E/\lambda}$ and the resulting equation reads

$$\frac{\partial W}{\partial t} = -\frac{\partial}{\partial E}\left[E\left(\tfrac{\dot{\lambda}}{2\lambda} - \gamma\right) + \gamma T\right]W + \gamma T\frac{\partial^2}{\partial E^2}(EW). \quad (4.21)$$

Note that we picked the algebraic sign $\pm$, so that in case of $\dot{\lambda} = 0$ the solution relaxes into the correct equilibrium distribution $\propto \exp(-E/T)$. Multiplying this equation by E and performing partial integrations on the right hand side we derive an evolution equation for the mean energy

$$\langle\dot{E}\rangle = \langle E\rangle\left(\tfrac{\dot{\lambda}}{2\lambda} - \gamma\right) + \gamma T. \quad (4.22)$$

For the variance in position we obtain

$$w_x = \langle x^2\rangle = \int x^2 W(x,E)dxdE = \int \frac{E}{\lambda}W(E)dE = \frac{\langle E\rangle}{\lambda}. \quad (4.23)$$

In addition we see from the definition of energy (4.13) and equation (4.23), that $w_v = \langle v^2\rangle = \langle E\rangle$, which means that the system energy is always equally distributed on position and velocity degrees of freedom which for quasi-static driving resembles the equipartition theorem from classical statistical mechanics for variables that enter the Hamiltonian quadratically [53]. The equation of motion for the variances then reads

$$\dot{w}_v = w_v\left(\tfrac{\dot{\lambda}}{2\lambda} - \gamma\right) + \gamma T, \qquad w_x = \frac{w_v}{\lambda}. \quad (4.24)$$

Note that the correlation between velocity and position vanishes in this limit

$$\langle xv \rangle = \int dx dE x \sqrt{2(E - \lambda x^2/2)} \frac{W(E)\sqrt{2\lambda}}{2\pi \sqrt{E - \lambda x^2/2}} = \int dE \frac{\sqrt{\lambda}}{\pi} W(E) \frac{1}{2} x^2 \Big|_{-\sqrt{\frac{2E}{\lambda}}}^{+\sqrt{\frac{2E}{\lambda}}} = 0. \qquad (4.25)$$

For a harmonic potential, the probability distribution in position and velocity is Gaussian. Hence, it is completely determined by solving the equations of motion (4.24) for the variances. The procedure to find the optimal functional form of the protocols is now as outlined in the last section. The irreversible work in this limit reads

$$W_{irr} = -2\gamma T(t_f - t_i) + \gamma \int_{t_i}^{t_f} \left[w_v + \frac{T^2}{w_v} \right] dt. \qquad (4.26)$$

Leading to an effective Lagrangian

$$\mathcal{L} = w_v + \frac{T^2}{w_v} + \alpha(t)\mathcal{C}, \qquad (4.27)$$

with the constraint derived from equation (4.24)

$$\mathcal{C} = w_v \left(\frac{\dot{\lambda}}{2\lambda} - \gamma \right) + \gamma T - \dot{w}_v = 0. \qquad (4.28)$$

Although the Lagrangian in equation (4.27) is simpler than the one in (4.10), the resulting nonlinear Euler-Lagrange differential equations still can not be solved analytically and also the numerical solution is out of scope due to the boundary value character of the problem[3].

4.3 Explicit Protocols

Since a successful search for the exact functional form of the optimal protocol is out of scope, we proceed by enforcing the shape of the protocol and subsequently optimize the free parameters to obtain a maximal power output. We systematically start our investigation with a simple linear protocol. In the experiment by Blickle and Bechinger described in section 3.1 a linear protocol was implemented and it is therefore definitely experimentally feasible. However, they did not optimize the protocol parameters. On the other hand, the group running the optical cavity setup described in section 3.2 assured us that they have very good control over the trapping laser, so that an optimization is possible. In a next step, we then subdivide the driving into a piecewise linear protocol and again optimize all free parameters to obtain a maximal power output. This can be understood as the second order of a systematic iteration procedure. Subdividing the protocol in infinitely many small linear sections, one can in principle approximate every possible functional form of the protocol and thereby come arbitrary close to the genuine shape of the optimal one. We will show in section 4.3.2 that second order already leads to very good results.

---

[3]The reasons are essentially the same ones as given at the end of section 4.1.

4.3.1 Simple Linear Protocol

A linear driving protocol generally has the following form

$$\lambda(t) : \begin{cases} \lambda_h(t) = c_{1h} + c_{2h}t, & \text{for } t < t_h, \\ \lambda_c(t) = c_{1c} + c_{2c}(t - t_h) & \text{for } t_h < t < t_h + t_c, \end{cases} \tag{4.29}$$

where t_h and t_c are the coupling times to hot and cold bath respectively. Note that this is the protocol just for one cycle which is then periodically repeated with period $t_{tot} = t_h + t_c$.

An important question is whether or not the protocol has to be continuous at the bath change and is discussed in detail in appendix A.4. We will here enforce continuity, which requires $\lambda_h(t_h) = \lambda_c(t_c)$ and $\lambda_h(0) = \lambda_c(t_c + t_h)$ and means that we can eliminate two of the constants in equation (4.29)

$$\lambda(t) : \begin{cases} \lambda_h(t) = \lambda_i - \dfrac{\lambda_i - \lambda_f}{t_h}t & \text{for } t < t_h, \\ \lambda_c(t) = \lambda_f + \dfrac{\lambda_i - \lambda_f}{t_c}(t - t_h) & \text{for } t_h < t < t_h + t_c. \end{cases} \tag{4.30}$$

For this protocol we can solve the equation of motion (4.23) for the mean square velocity analytically

$$w_v(t) = e^{\gamma(f - t)}\sqrt{f - t}\left[c_I - \sqrt{\gamma \pi}T\,\mathrm{erf}(\sqrt{\gamma(f - t)})\right] \quad \text{with:} \tag{4.31}$$

$$f = \begin{cases} \frac{t_h}{(1-b)} \\ t_h - \frac{bt_c}{1-b} \end{cases}, \quad T = \begin{cases} T_h \\ T_c \end{cases}, \quad \gamma = \begin{cases} \gamma_h \\ \gamma_c \end{cases}, \quad c_I = \begin{cases} C_h & \text{for } t < t_h \\ C_c & \text{for } t_h < t < t_h + t_c \end{cases}, \tag{4.32}$$

where we have introduced the ratio between minimal and maximal trapping stiffness $b = \lambda_f/\lambda_i$. Note that w_v only depends on the ratio b and not on λ_i or λ_f independently. The integration constants c_I are determined by demanding continuity of w_v at the bath change and result to be

$$C_h = \frac{\sqrt{\frac{\pi}{t_h}}e^{\frac{2\gamma_c t_c}{1-b}}}{e^{\gamma_c t_c + \gamma_h t_h} - 1}\left[T_h\sqrt{\gamma_h t_h}\left(e^{\gamma_h t_h - \frac{(b+1)\gamma_c t_c}{1-b}}\mathrm{erf}\left(\sqrt{\frac{\gamma_h t_h}{1-b}}\right) - e^{\frac{2\gamma_c t_c}{b-1}}\mathrm{erf}\left(\sqrt{\frac{b\gamma_h t_h}{1-b}}\right)\right) \right.$$
$$\left. + T_c\sqrt{\gamma_c t_c}\left(\mathrm{erfi}\left(\sqrt{\frac{\gamma_c t_c}{1-b}}\right) - \mathrm{erfi}\left(\sqrt{\frac{b\gamma_c t_c}{1-b}}\right)\right)e^{-\frac{(b+2)\gamma_c t_c + b\gamma_h t_h}{1-b}}\right], \tag{4.33}$$

$$C_c = \frac{i\sqrt{\pi}}{\sqrt{t_c}\left(e^{\gamma_c t_c + \gamma_h t_h} - 1\right)}\left[T_h\sqrt{\gamma_h t_h}e^{\frac{\gamma_c t_c + \gamma_h t_h}{1-b}}\left(\mathrm{erf}\left(\sqrt{\frac{\gamma_h t_h}{1-b}}\right) - \mathrm{erf}\left(\sqrt{\frac{b\gamma_h t_h}{1-b}}\right)\right) \right.$$
$$\left. + T_c\sqrt{\gamma_c t_c}\left(\mathrm{erfi}\left(\sqrt{\frac{\gamma_c t_c}{1-b}}\right) - e^{\gamma_c t_c + \gamma_h t_h}\mathrm{erfi}\left(\sqrt{\frac{b\gamma_c t_c}{1-b}}\right)\right)\right]. \tag{4.34}$$

Here we have assumed that the two temperature baths T_h and T_c are associated with two different damping constants γ_h and γ_c. Therefore the solutions are also applicable to a heat engine run by optical damping where effective damping constants and temperatures are given by relations (3.10).

Optimization of Parameters

Our goal is to optimize the parameters b, t_h and t_c to maximize the power output of the heat engine. Since the process is cyclic, the net change in energy over one cycle is zero and with the help of the energy balance equation (1.45) and the expression for the average heat transfer (1.57) we can express work and power output as

$$-\Delta W = \Delta Q_h + \Delta Q_c$$

$$= \gamma_h \left(T_h t_h - \int_0^{t_h} w_v dt \right) + \gamma_c \left(T_c t_c - \int_{t_h}^{t_h+t_c} w_v dt \right), \tag{4.35}$$

$$P = -\Delta W/(t_h + t_c). \tag{4.36}$$

The efficiency is then obtained according to

$$\eta = -\frac{\Delta W}{\Delta Q_h} = 1 - \frac{\gamma_c}{\gamma_h} \frac{T_h t_h - \int_0^{t_h} w_v dt}{T_c t_c - \int_{t_h}^{t_h+t_c} w_v dt}. \tag{4.37}$$

All integrals in these expressions are of the form

$$\int_{t_i}^{t_f} w_v(t)dt = \frac{1}{2\gamma^{3/2}} \left[-2\gamma T \sqrt{f - t_i} \sqrt{\gamma(f - t_i)} \, {}_2F_2 \left(1, 1; \frac{1}{2}, 2; \gamma(f - t_i) \right) \right.$$

$$+ 2\gamma T \sqrt{f - t_f} \sqrt{\gamma(f - t_f)} \, {}_2F_2 \left(1, 1; \frac{1}{2}, 2; \gamma(f - t_f) \right) + c_1 \left(\sqrt{\pi} \left(\text{erfi} \left(\sqrt{\gamma} \sqrt{f - t_f} \right) \right. \right.$$

$$\left. - \text{erfi} \left(\sqrt{\gamma} \sqrt{f - t_i} \right) \right) - 2\sqrt{\gamma} \left(\sqrt{f - t_f} e^{\gamma(f - t_f)} - \sqrt{f - t_i} e^{\gamma(f - t_i)} \right) \right)$$

$$\left. + 2\gamma T \left(\sqrt{f - t_i} \sqrt{\gamma(f - t_i)} - \sqrt{f - t_f} \sqrt{\gamma(f - t_f)} \right) \right], \tag{4.38}$$

where the c_1, T and γ have to be chosen according to the time interval and the ${}_pF_q(a; b; z)$ denote generalized hyper-geometric functions [54]. This means that we have analytical expressions at hand for the power P and efficiency η. These quantities show highly nonlinear dependencies on the parameters b, t_h and t_c, but can be optimized numerically for specific values of damping and temperature.

In the following we substitute $t_{tot} = t_c + t_h$ and $t_c = a t_h$ and maximize the power output of the engine numerically with respect to the parameters a and b for varying total cycle times t_{tot}. Generally this can be done for arbitrary parameter sets. Here we exemplarily choose temperatures of

$$T_h = 4, \quad T_c = 1, \tag{4.39}$$

and show results for the case of real temperature baths with constant damping γ_{th} and for the case of an effective temperature change via optical damping, where after fixing T_h and T_c the damping constant are given by relations (3.10). A factor of 4 in the temperature difference is experimentally accessible. With the technique of optical damping our experimentalist colleagues in Vienna can even achieve an effective temperature difference up to a factor of 10. Note that we are working with rescaled units as introduced in equation (A.2).

In figure 4.1 the optimized power output is shown for three different damping constants, $\gamma_{th} = 0.001$ (4.1(a)), $\gamma_{th} = 0.01$ (4.1(b)) and $\gamma_{th} = 0.1$ (4.1(c)). We see clear maxima at optimal coupling times t^*_{tot} that depend on γ_{th}, which determines the timescale of the process. The existence of an optimal coupling time for maximal power output can be explained as follows. The heat exchange between temperature bath and system takes place through the kinetic degrees of freedom. For very short coupling times, the mean square velocity does not have time to relax to the thermal velocity nor to a value considerably higher than during the cold phase. However, this is essential for the engine to perform work and consequently the power output goes to zero. On the other hand, the power output becomes small for very long cycle times because it is defined as work output divided by the coupling time. In the case of effective temperatures via optical damping, the power output is generally higher and the maximum is shifted to smaller times. This is due to the fact that the damping constant during the cold phase is increased, which means stronger coupling to the bath and therefore a faster timescale on which heat exchange takes place.

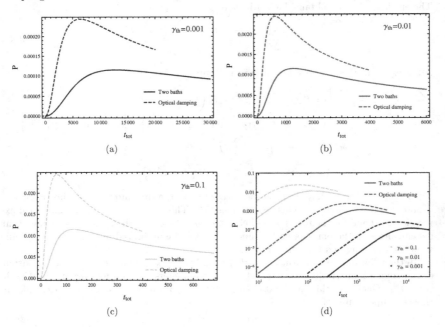

Figure 4.1: Power curves plotted over the total cycle time for different damping constants γ_{th} (a)–(c). The curves have a clear maximum at some optimal coupling time t^*_{tot}. (d) displays the log-log plot of (a)–(c). We see that the shape of the curves for different γ_{th} is very similar, only the timescale of the process is shifted by the damping.

In figure 4.2(a) the optimized parameter a^* is depicted to obtain maximum power output for a fixed cycle time corresponding to the power curves from figure 4.1. In the case of real temperature baths we see that a^*, the ratio of coupling times to the hot and cold heat bath respectively, increases with increasing cycle time t_{tot}. Generally the optimal coupling time to the cold bath is longer than to the hot bath. Interestingly this is different for the case of optical damping. Asymptotically for small and for large cycle times the ratio of coupling times approaches unity, but generally the system couples longer to the hot than to the cold bath. This is explained by the fact that the effective damping at the cold bath is higher, meaning a stronger coupling to the bath. The process at the cold bath therefore takes place on a faster timescale than in the case where the temperature difference is real and the damping remains unchanged.

Figure 4.2(b) shows the optimized parameter b^* that gives the ratio between initial and final trap stiffness and which is a monotonically decreasing function of the cycle time t_{tot}. For short cycle times b^* is very close to one. This is due to the fact that little heat is exchanged during the short coupling times and changing the trap stiffness rapidly would require to put work into the system instead of extracting work from the system. For very long cycle times on the other hand, b^* goes to zero. This can be explained as follows. As long as the system is not in equilibrium, heat will flow between bath and system. This mechanism is used to extract work. If the cycle time is increased without decreasing b^* the system would remain close to equilibrium where heat exchange is very low and therefore hardly any work could be extracted during the cycle. No qualitative difference between real and effective heat engine occurs, only the timescale is slightly shifted due to the different damping at the cold phase.

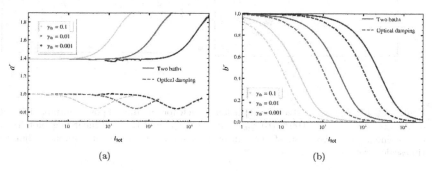

(a)   (b)

Figure 4.2: Optimized free parameters to obtain maximum power output for fixed total cycle time. (a) The optimal ratio of coupling times $a^* = t_c/t_h$ over cycle time is depicted. (b) Optimal ratio of minimal and maximal trap stiffness $b^* = \lambda_i/\lambda_f$ over cycle duration is shown.

Figure 4.3 shows the mean efficiencies η over the total cycle time t_{tot} corresponding to a protocol with the optimized parameters a^* and b^* from figure 4.2 to maximize the power output. We see that η is a monotonically increasing function of t_{tot}. The efficiency in case of optical damping is higher than in the case of a normal heat engine, but for both cases it is of course bounded from above by the Carnot efficiency. Carnot efficiency could only be achieved when all processes were performed at quasi-equilibrium and therefore

reversibly. Since we are dealing with finite time processes, we have a work loss due to
entropy production (see equation (1.64)). Different thermal damping constants γ_{th} shift
the timescale but do not have an effect on the functional dependence of η on t_{tot}.

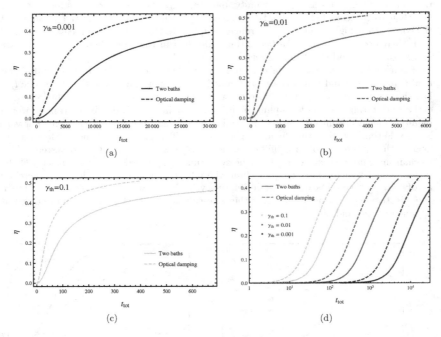

Figure 4.3: Efficiencies plotted over duration of cycle for maximized power output. Thermal
dampings of (a) $\gamma_{th} = 0.001$, (b) $\gamma_{th} = 0.1$, (c) $\gamma_{th} = 0.1$. (d) Linear-Log plot for a–
c combined. The evolution of the efficiency is similar for different dampings, but the
timescale strongly depends on γ_{th}.

In figure 4.4 the time evolution of the mean squared velocity w_v and the mean squared
position w_x is shown exemplarily for $\gamma_{th} = 0.01$ at optimized parameters t_{tot}^*, b^* and a^*
that give maximum power. The corresponding protocol for the control parameter $\lambda(t)$ is
depicted in figure 4.4(c). We here compare the approximate analytical solution with exact
numerical results to see whether the approximation is justified. The analytical solution
for w_v is independent of the initial trapping stiffness λ_i and only depends on the ratio
$b = \lambda_f/\lambda_i$ between minimal and maximal control parameter. This is not entirely true for
the exact solutions, since λ determines the frequency of oscillations of particles in the trap.
These have been eliminated in the approximation by assuming that the energy was a slowly
changing variable. Generally we see from figure 4.4 that the numerical solutions oscillate
around the analytical solution but on average agree very well. Note that we are interested

in quantities like work, power and efficiency, whose mean values depend on the integral of w_v over time. If the frequency of oscillation is high, these integrals will differ hardly from the approximate analytical solution, which is thus best for high trapping frequencies.

In figure 4.5(a) and (b) the efficiency and power are plotted over the initial trap stiffness λ_i. As expected, for high λ_i the analytical approximation coincides with the numerical exact result; whereas for very small λ_i they differ. This exactly reflects the assumption that the energy of the system does not vary much over one period of oscillation. In the light of experimental reproduction of the analytical results, it is therefore favorable to work with high trapping frequencies.

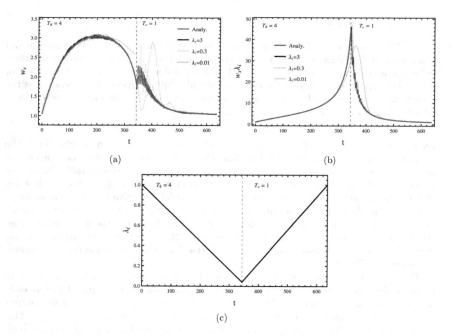

(a)

(b)

(c)

Figure 4.4: Second moments of velocity w_v and position w_x are shown for one cycle at maximum power with optical damping and $\gamma_{th} = 0.01$. Comparison of the analytical approximate solution to the exact numerical solutions for different initial trapping stiffnesses.(a) The analytical approximate solution for w_v is independent of λ_i, whereas the oscillation frequencies of the exact solutions strongly depend on it. (b) For reasons of comparability w_x is multiplied by λ_i. The analytical approximation is most accurate for large λ_i since the oscillation frequency is higher and so the energy changes less during one oscillation.(c) The protocol of the control parameter $\lambda(t)$ corresponding to plots (a),(b).

Figure 4.6(a) shows the distributions of the entropy that is dissipated into the envi-

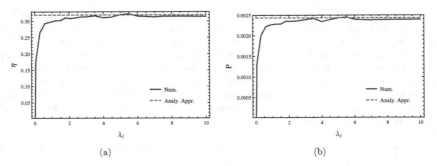

(a)   (b)

Figure 4.5: Comparision of (a) efficiency and (b) power of the analytic approximation and exact numerical solution for different values of the initial trap stiffness λ_i. The temperature difference is created via optical damping. The analytical approximation is independent of the initial trap stiffness whereas the exact numerical solution depends on λ_i as soon as the energy changes significantly during one oscillation. Parameter values are $\gamma_{th} = 0.01, a = 0.85, b = 0.041, T_h = 4, T_c = 1$.

ronment during one cycle for different values of λ_i. Only for very small values of λ_i the distributions differ. This indicates that also the dissipated entropy does not depend on the absolute scale of λ_i, unless the energy of the system changes considerably during one period of oscillation. Note that on average the second law of thermodynamics, which states that the total entropy production is always positive, is fulfilled. Due to fluctuations, single trajectories might still follow a path along which entropy production is negative. Therefore there is a nonzero probability for negative S_{env}. In figure 4.6(b) the work distributions are depicted. The mean values are negative meaning that on average the engine performs work. Note the characteristic non-Gaussian tail, that has been observed in driven harmonic potentials [55, 56, 57].

Finally, in addition to a and b we also treat the cycle time t_{tot} as a free parameter to optimize the power output. As already seen by the clear maxima in figure 4.1 there exits an optimal total cycle time t_{tot}^*. Figure 4.7(a) shows the maximal possible power output over the thermal damping, which apparently increases linearly with increasing γ_{th}. This makes sense, since the damping constant γ_{th} is a measure for the coupling between bath and system.

Figure 4.7(b) shows the important efficiency at maximum power which, as in the overdamped case, is constant and therefore independent of γ_{th}. Compared to analytical results in the overdamped limit it is significantly smaller, for our parameter set it is 0.278 (overdamped 0.462). In the case of optical damping we obtained a slightly higher efficiency at maximum power of 0.319 which resembles the trend in the overdamped limit (0.428). An explanation of the lower efficiencies is the fact that we enforced the form of the protocol to be linear. For the case of a piecewise linear protocol the efficiencies are already much closer to the ones obtained in the overdamped limit (see section 4.3.2).

Figure 4.7(c) and figure 4.7(d) show the optimal parameters a^* and b^* which likewise are

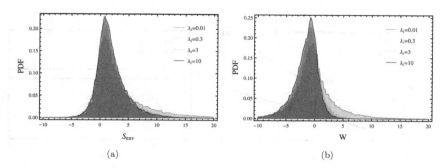

(a)   (b)

Figure 4.6: (a) Probability distribution function for the dissipated entropy and (b) for the performed work during one cycle of a heat engine operated at maximum power via optical damping for $\gamma_{th} = 0.01$, $T_h = 4$, $T_c = 1$. For different λ_i's the distributions differ only if λ_i becomes small enough so that the system's energy changes considerably over one oscillation period. The mean work is negative which implies that the engine on average performs work. The mean entropy production is positive, so that on average the 2nd law is fulfilled.

constant and do not depend on γ_{th}. Note that in the case where the effective temperature difference is achieved via optical damping, the system couples longer to the hot bath and in the case of real temperatures the system couples longer to the cold bath.

Figure 4.7(e) shows the dependence of the optimal cycle duration t^*_{tot} on the thermal damping γ_{th}. Note that the scale is double logarithmic. We clearly see that the timescale of optimal coupling is determined by the damping, i.e. $t^*_{tot} = c/\gamma_{th}$, where the constant c depends on the bath temperatures and on the protocol parameters. It therefore is proportional to the timescale of velocity relaxation τ_{rel}. This is completely different from the dependence between optimal coupling time and damping in the overdamped case where $t^*_{tot} \propto \gamma_{th}$ (see equation (2.20)). It therefore makes perfect sense that the optimal protocols derived in the overdamped limit fail in the underdamped regime as we already showed in section 2.3.2.

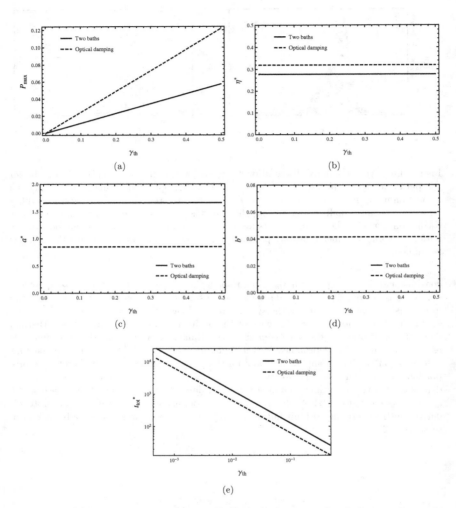

Figure 4.7: (a) Maximum power output of a linearly driven stochastic heat engine under optimized parameters a^*, b^*, t^*_{tot} for varying thermal damping γ_{th}. (b) Efficiency at maximum power over γ_{th}. (c) Optimized parameter a^* for maximal power output over γ_{th}. (d) Optimized parameter b^* for maximal power output over γ_{th}. (e) Log-Log plot of total cycle time versus γ_{th}.

4.3.2 Piecewise Linear Protocol

As we have seen in section 4.3.1 the equations of motion (4.24) for the variances in case of a linear potential can be solved analytically. Naturally, the next step to optimize the heat engine further is to consider a piecewise linear potential where the nodes are placed at half the coupling times to the hot and cold baths. Enforcing continuity, the protocol reads

$$
\lambda(t) = \begin{cases}
\lambda_i \left(1 + \dfrac{c_h - 1}{t_h/2} t \right) & \text{for } t < \frac{t_h}{2}, \\[2mm]
\lambda_i \left(c_h + \dfrac{(b - c_h)}{t_h/2}(t - t_h/2) \right) & \text{for } \frac{t_h}{2} < t < t_h, \\[2mm]
\lambda_i \left(b + \dfrac{c_c - b}{t_c/2}(t - t_h) \right) & \text{for } t_h < t < t_h + \frac{t_c}{2}, \\[2mm]
\lambda_i \left(c_c + \dfrac{1 - c_c}{t_c/2}(t - t_h - t_c/2) \right) & \text{for } t_h + \frac{t_c}{2} < t < t_h + t_c,
\end{cases}
\tag{4.40}
$$

where c_h, b and c_c are treated as free parameters. A diagrammatic sketch of the potential with the free parameters included is shown in figure 4.8. The solution of the equations of motion (4.24) then reads

$$
w_v(t) = e^{\gamma(f-t)} \sqrt{f - t} \left[c_I - \sqrt{\gamma \pi} T \, \mathrm{erf}(\sqrt{\gamma(f-t)}) \right] \quad \text{with:}
\tag{4.41}
$$

$$
f = \begin{cases}
\dfrac{t_{tot}}{2(1+a)(1-c_h)} \\[2mm]
\dfrac{(c_h - b/2)t_{tot}}{(1+a)(c_h - b)} \\[2mm]
\dfrac{(b - c_c + ab/2)t_{tot}}{(1+a)(b - c_c)} \\[2mm]
\dfrac{(c_c - 1 + a(c_c - 1/2))t_{tot}}{(1+a)(c_c - 1)}
\end{cases}
, T = \begin{cases}
T_h \\ T_h \\ T_c \\ T_c
\end{cases}
, \gamma = \begin{cases}
\gamma_h \\ \gamma_h \\ \gamma_c \\ \gamma_c
\end{cases}
, c_I = \begin{cases}
C_{h1} & \text{for } t < \frac{t_h}{2} \\
C_{h2} & \text{for } \frac{t_h}{2} < t < t_h \\
C_{c1} & \text{for } t_h < t < t_h + \frac{t_c}{2} \\
C_{c2} & \text{for } t_h + \frac{t_c}{2} < t < t_h + t_c
\end{cases}
,
$$

$$
\tag{4.42}
$$

where the integration constants C_{h1}, C_{h2}, C_{c1} and C_{c2} are determined through the continuity condition on w_v and can be found in the appendix A.5 (equations (A.43)–(A.46)).

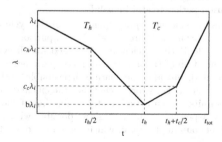

Figure 4.8: Scematic peacewise-linear driving protocol $\lambda(t)$ for one cycle.

Optimization of Parameters

Analogous to section 4.3.1 we proceed by optimizing the power with respect to the free parameters. The expressions for work, power and efficiency are given by equations (4.36) and (4.37), where the occurring integrals (4.38) have to be evaluated with the help of equation (4.42).

Figure 4.9(a) shows the power curves over the total cycle times for real temperature baths and for effective temperature differences generated via optical damping. Here all free parameters have been optimized to achieve a maximum power output. We see two main differences compared to the previous section with a simple linear protocol (figure 4.1). First, the maximal power output is increased significantly by nearly a factor of 2, due to the optimization of the additional degrees of freedom c_h and c_c in the protocol. Secondly, at small cycle times it is now possible to achieve a high power output, where for the simple linear protocol from section 4.3.1 the power output went down to zero. How this difference comes about is best explained by looking at a concrete example. Figure 4.10(f) shows the optimized protocol for a short cycle time of $t = 20$ for a parameter set of $T_h = 4, T_c = 1, \gamma_{th} = 0.01$ and the temperature difference created via optical damping. Due to the additional degree of freedom c_h in the protocol, it is possible to first increase the trapping stiffness. This leads to a faster increase of the mean square velocity, which can be observed in figure 4.10(b). The subsequent opening of the trapping potential is then performed when the particles have already picked up thermal energy. The net change of the trapping potential during the coupling to the hot bath is thereby negative, which can be seen in figure 4.9(f), i.e. the parameter b remains smaller than 1. Figures 4.10(a) and 4.10(c) show the second moments of velocity and position for the same parameter set at maximum power. Figure 4.10(e) depicts the corresponding protocol for the trap stiffness. In contrast to the example of the optimized protocol for a short cycle time of $t_{tot} = 20$ the optimal values for c_h and c_c now lie between b and 1, corresponding to opening the trap when coupling to the hot bath and closing the trap when coupling to the cold bath and thereby showing resemblance to the simple linear protocol[4]. Generally, we see from figure 4.10 that the analytical approximation becomes the more accurate, the bigger the scale of the trapping stiffness λ_i, as already discovered in the last section. Note that especially at short coupling times the trapping stiffness changes rapidly. In this case, a very high initial trapping stiffness λ_i is required for the approximation's assumption of slowly varying energy over an oscillation period to still be valid.

Figure 4.9(b) shows the efficiency η in dependence of the cycle time t_{tot} corresponding to the optimized power output in figure 4.9(a). In comparison to the curve for a simple linear protocol (figure 4.3(b)) the efficiency is increased for all cycle times. This means that through the optimization of the newly introduced parameters c_h and c_c not only the extracted work is increased, but also the share of irreversible work lost by dissipation in the exchanged heat is reduced. For short cycle times the efficiency does not fall off to zero anymore, but remains finite. In an experimental realization of such a heat engine, one is interested in obtaining a power output as high as possible with a high efficiency at the same time. The range of coupling times, in which these requirements are met, is expanded

---

[4]The simple linear protocol corresponds to parameters $c_h = c_c = (1 + b)/2$.

considerably through the introduction of the parameters c_h and c_c and now reaches from small cycle times to cycle times where the actual maximum of power is encountered.

The optimized parameters a^*, b^*, c_h^* and c_c^* are depicted in figures 4.9(c)–4.9(f). Several remarks are in order. The ratio of coupling times to cold and hot baths a^* is decreased compared to the case of a simple linear protocol (figure 4.2(a)) and also the functional dependence is altered. Concerning the dependence of the optimized parameters c_h^*, b^* and c_c^* on t_{tot} (figure 4.9(e)), we clearly see the transition between three qualitatively different regimes of the protocol, which in the case of a simple linear protocol was not possible.

1. For very short cycle times, the power is maximized by a protocol characterized by $c_h^* > 1 > b^* > c_c^*$. This means that during the coupling to the hot bath the trapping stiffness is first increased and then decreased, with a negative net change in λ. During the coupling to the cold bath the trap stiffness is first decreased and then increased up to the initial trapping stiffness.

2. The second regime is characterized by $c_h^* > 1 > c_c^* > b^*$. The difference to the former regime is, that during the coupling to the cold bath the stiffness increases monotonously, but with different slopes during the first and second half of the coupling time.

3. Finally in the third regime, characterized by $c_h^* > 1 > c_c^* > b^*$, the trap stiffness is decreased during the coupling to the hot bath and increased during the coupling to the cold bath.

Subsequently, we also treat the total cycle time t_{tot} as a free parameter and numerically find the maximal power output for different values of the damping constant γ_{th}. We have seen exemplarily for $\gamma_{th} = 0.01$ that a maximum exists (see figure 4.9(a)), although it is not as pronounced anymore as it was in the case of a simple linear protocol. The findings are depicted in figure 4.11(a). As in section 4.3.1 the maximal power shows a linear dependence on the thermal damping constant γ_{th}. Note that the slope is considerably higher compared to the simple linear protocol. This shows that the subdivision of the linear protocol and the optimization of the parameters c_h and c_c indeed improved the power output of the engine.

Figure 4.11(b) shows the efficiency at maximum power, which is independent of the thermal damping γ_{th}. Compared to the simple linear protocol it is considerably higher, namely 0.376 (with optical damping 0.395) compared to 0.278 (0.319) and much closer to the value of 0.462 (0.428) obtained with the current parameter set for an optimal protocol in the overdamped limit. This indicates that the efficiency at maximum power might be the same for the optimal protocols in over-, and underdamped regimes, but for conclusive evidence one would have to investigate protocols with even more subdivisions to imitate more accurately the genuine optimal protocol. In any way the efficiency remains smaller than the Curzon-Ahlborn efficiency predicted for a classical heat engine at maximum power.

For completeness we also show the graphs of the optimized free parameters. Figure 4.11(c) displays the ratio of coupling times a^* over the thermal damping constant. The optimal parameter a^* turns out to be independent of γ_{th}. Compared to the simple linear protocol, a^* is decreased for both cases of real temperature baths and optical damping. This means that the coupling time to the cold bath is shortened through the subdivision of the protocol.

Figure 4.11(d) depicts the three optimization parameters of the protocol c_h^*, b^*, c_c^* over

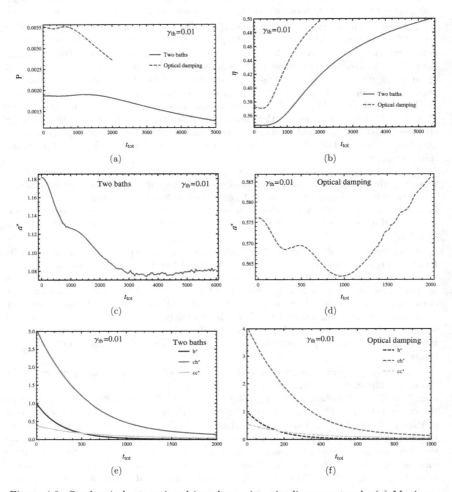

Figure 4.9: Stochastic heat engine driven by a piecewise linear protocol. (a) Maximum power output under optimized parameters a^*, b^*, c_h^*, c_c^* for varying total cycle time t_{tot} and (b) the corresponding efficiencies. (c,d) Optimized ratio of coupling times to cold and hot bath a^* corresponding to maximized power output. (e,f) Optimized parameter b^*, c_h^* and c_c^* corresponding to maximized power output.

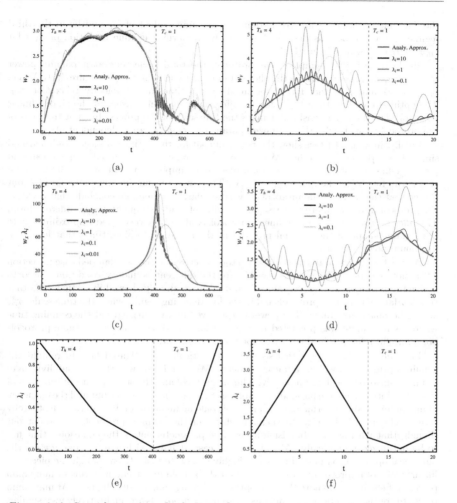

Figure 4.10: Second moments of velocity and position for one cycle of a heat engine run by optical damping, exemplarily shown for the cycle time which gives maximum power (a,c) and a shorter cycle time of $t_{tot} = 20$ (b,d) at a thermal damping of $\gamma_{th} = 0.01$. (e,f) Show the according protocol of the control parameter λ with optimized parameters c_h^*, b^*, c_c^*, a^*. In (a,b) the analytical approximate solution for w_v is independent of λ_i, whereas the oscillation frequency of the exact solutions strongly depends on it. (c,d) For reasons of comparability the variance in position w_x is multiplied by the initial stiffness λ_i. The analytical approximation is most accurate for large λ_i since the oscillation frequency is higher and so the energy changes less during one oscillation, which was part of the assumption to derive the analytical result.

the damping constant. All three parameters are independent of γ_{th} and lie in the third regime of the protocol as described above. Exemplarily the time evolution of $\lambda(t)$ for $\gamma_{th} = 0.01$ is shown in figure 4.10(e).

Similarly as for the simple linear protocol, the total cycle time at maximum power t_{tot}^* is inversely proportional to the thermal damping constant γ_{th} (see figure 4.11(e)), i.e. $t_{tot}^* = c/\gamma_{th}$ and it is therefore determined by the timescale of velocity relaxation τ_{rel}. The optimal coupling time t_{tot}^* is hardly affected by the subdivision of the simple linear protocol. The damping constant changes the timescale of the process but not the form of the protocol. This is why the optimized parameters c_h^*, b^*, c_c^* are independent of γ_{th}.

Finally, in figure 4.12 we show the power output for the optimized piecewise-linear and simple linear protocols together. We exemplarily compare the optimized simple linear and piecewise-linear protocols at maximum power for a damping constant of $\gamma_{th} = 0.01$ and an engine run by optical damping (figure 4.13). The power output differs by a factor of 2, but we see that the shape of the protocols is very similar. We therefore conclude that dividing the protocol into even more linear segments will not lead to a qualitatively different form. This means, that the piecewise-linear protocol is already a very good approximation to the genuine shape of the optimal protocol in the range of coupling times that lead to a maximum power output.

A direct comparison to the optimal protocols derived in the overdamped case in section 2.2 is not possible. The reasons for this are the different boundary conditions[5] used to derive the protocols in the overdamped limit. But we already showed in section 2.3.2 that the overdamped optimal protocols fail in the low damping regime when the time scales τ_{rel} and τ_λ become comparable. The power output even became negative at the coupling time where its maximum was predicted by the overdamped theory. Therefore these protocols are no useful reference in the unerdamped regime, anyway.

In summary, we have in this section systematically investigated how to rebuild the genuine optimal protocol from linear segments. We thereby showed that the subdivision of a simple linear protocol into a piecewise linear protocol and the subsequent optimization of the newly introduced free parameters improves the maximal power output and the efficiency at maximum power. In principle this protocol could again be subdivided iteratively, thereby coming closer and closer to the genuine shape of the optimal protocol. We point out though, that with every subdivision a new free parameter enters the equations, that has to be optimized numerically. Already for the case at hand with five free parameters the numerical optimization procedure was a highly elaborate task. Even though the piecewise linear protocol improves quantitatively upon the simple linear one in terms of maximum power and efficiency, we note that the optimal coupling time hardly changes. At maximum power the optimal simple linear and piecewise-linear protocols are qualitatively very similar.

---

[5]In the overdamped limit the boundary conditions are imposed on the mean squared position, i.e. w_a and w_b. Although these can be translated back into a protocol λ with equation (2.15), it still depends on the imposed values w_a and w_b and besides exhibits jumps. A scenario for a possible comparison at maximum power would be to fix λ_i and consequently $\lambda_f = b^* \lambda_i$ and drive the protocol from λ_i to λ_f at the hot bath and from λ_f to λ_i at the cold bath. Then we could use equation (2.15) to recalculate the boundary values for the variances in the overdamped case. This on the other hand leads to discontinuities in the variances w_a and w_b at the bath change, which is unphysical and runs against the arguments used to derive the protocols in this limit.

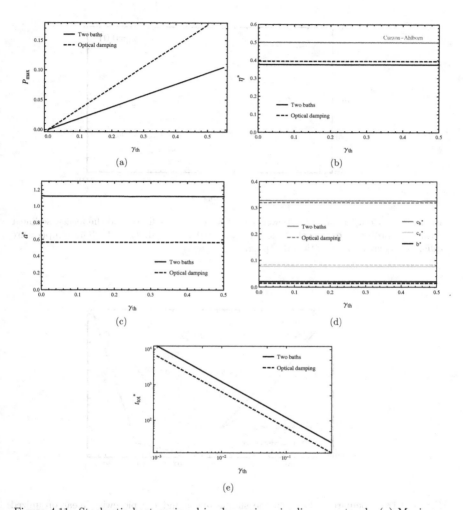

Figure 4.11: Stochastic heat engine drive by a piecewise linear protocol. (a) Maximum power output under optimized parameters $a^*, c_h^*, b^*, c_c^*, t_{tot}^*$ for varying thermal damping γ_{th}. (b) Efficiency at maximum power over γ_{th}. (c) Optimized parameter a^* for maximal power output over γ_{th}. (d) Optimized parameters c_h^*, b^*, c_c^* for maximal power output over γ_{th}. (e) Log-Log plot of total cycle time versus γ_{th}.

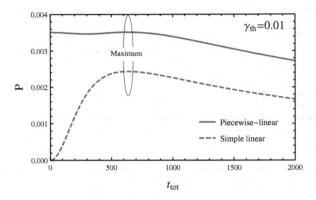

Figure 4.12: Comparison of the power output for the simple linear and the piecewise-linear optimized protocols over cycle time for $\gamma_{th} = 0.01$, $T_h = 4$, $T_c = 1$. The temperature difference is created via optical damping.

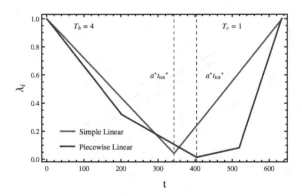

Figure 4.13: Comparison between the simple linear and the piecewise-linear optimized protocols at maximum power for $\gamma_{th} = 0.01$, $T_h = 4$, $T_c = 1$. Temperature difference is created via optical damping. The total optimal cycle time is nearly the same for both. The coupling time to the hot bath is slightly longer for the piecewise-linear protocol. Qualitatively, both protocols are very similar.

This suggests that a further subdivision of the protocol is unlikely to lead to a qualitative difference at optimal coupling times. We therefore conclude, that the optimized piecewise-linear protocol is indeed very close to the genuine optimal protocol.

5

Entropy Production in Inhomogeneous Thermal Environments

In chapters 2–4 we have investigated a heat engine in the overdamped and underdamped regime. We thereby saw that entropy production in the thermal environment and the associated irreversible work are crucial for the resulting efficiency and power output of the engine. So far we did not investigate possible temperature gradients which influence entropy production severely.

It has been shown recently [24] that the overdamped Langevin equation does not lead to the correct limiting statistics of entropy production in the high-friction limit when temperature gradients are present in the system. In the following, we will first present the theory developed in [24]. Then we will investigate an analytically accessible example system and finally perform a numerical study for the optical trap setup described in chapter 3.2.

5.1 Entropic Anomaly

Equation (1.54) gives the entropy dissipated into the environment along a single trajectory. In n dimensions it reads

$$s_{env} = -\int_{t_0}^{t} \frac{1}{T} \left(-f^i v^i + v^i \dot{v}^i \right) d\tau, \quad i = 1, ..., n. \tag{5.1}$$

With $\omega = \exp(-\frac{v^i v^i}{2T})/(2\pi T)^{n/2}$ this can be rewritten as

$$
\begin{aligned}
s_{env} &= \int_{t_0}^{t} \left[\frac{f^i}{T} v^i - \frac{v^i}{T} \frac{dv^i}{d\tau} - \frac{d}{d\tau} \ln \omega \right] d\tau + \ln \left(\frac{\omega_t}{\omega_{t_0}} \right) \\
&= \int_{t_0}^{t} \left[\frac{f^i}{T} v^i - \frac{v^i}{T} \frac{dv^i}{d\tau} + \frac{d}{d\tau} \left(\frac{v^j v^j}{2T} + \frac{n}{2} \ln T \right) \right] d\tau + \ln \left(\frac{\omega_t}{\omega_{t_0}} \right) \\
&= \int_{t_0}^{t} \left[\frac{f^i}{T} v^i + \frac{nT - v^j v^j}{2T} \frac{d}{d\tau} \ln T \right] d\tau + \ln \left(\frac{\omega_t}{\omega_{t_0}} \right) \\
&= \int_{t_0}^{t} \left[\frac{f^i}{T} v^i + \frac{nT - v^j v^j}{2T^2} \left(\frac{\partial T}{\partial \tau} + v^i \frac{\partial T}{\partial x^i} \right) \right] d\tau + \ln \left(\frac{\omega_t}{\omega_{t_0}} \right),
\end{aligned}
\tag{5.2}
$$

and the integral can furthermore be split in three different parts

$$s_{env} = \int_{t_0}^{t} \left[\underbrace{\left(\frac{f^i}{T} - \frac{1}{T} \frac{\partial T}{\partial x^i} \right) v^i}_{s_{reg}} + \underbrace{\frac{nT - v^i v^i}{2T^2} \frac{\partial T}{\partial \tau}}_{s_{time}} + \underbrace{\frac{(n+2)Tv^i - v^j v^j v^i}{2T^2} \frac{\partial T}{\partial x^i}}_{s_{anom}} \right] d\tau + \ln \left(\frac{\omega_t}{\omega_{t_0}} \right). \quad (5.3)$$

The entropy produced in the environment is a measure of irreversibility of the process. It can be connected to the ratio of probabilities for the forward and backward trajectory under time reversal [58, 59]

$$\log \left(\frac{\mathcal{P}^{\mathcal{F}}[X]}{\mathcal{P}^{\mathcal{R}}[\bar{X}]} \right) = s_{env}, \quad (5.4)$$

where $\mathcal{P}^{\mathcal{F}}[X]$ is the probability of the forward trajectory and $\mathcal{P}^{\mathcal{R}}[\bar{X}]$ the probability for the reversed trajectory. Using short-time propagators and the path integral formalism it can be shown[24, 60], that the part denoted as s_{reg} in equation (5.3) is exactly the entropy production of an overdamped trajectory described by the overdamped Langevin equation

$$dx^i = \left(\frac{f^i}{\gamma} - \frac{1}{2\gamma} \frac{\partial T}{\partial x^i} + \frac{T}{2} \frac{\partial \gamma^{-1}}{\partial x^i} \right) dt + \sqrt{2T/\gamma} dW^i. \quad (5.5)$$

The part s_{time} is the entropy production associated with a time-dependent temperature and s_{anom} will be called anomalous entropy production. The name is due to the fact, that it can not be associated with a functional over overdamped trajectories and therefore is not modeled by the overdamped Langevin equation (5.5).

We want to derive the high-friction limit for the entropy production in equation (5.3). The joint generating function for the different parts in the integral reads

$$G_c(x, v, t | x', v', t') = \langle \exp(-c_1 s_{reg} - c_2 s_{time} - c_3 s_{anom}) \delta(X_t - x) \delta(V_t - v) \rangle, \quad (5.6)$$

where the average is taken over paths with fixed initial and final configurations and (X_t, V_t) is the time-dependent stochastic trajectory. G_c fulfills a forward Feynman-Kac formula (see appendix A.6), which is of the form

$$\left(\frac{\partial}{\partial t} - L^\dagger \right) G_c = - \left[c_1 \left(\frac{f^i}{T} - \frac{1}{T} \frac{\partial T}{\partial x^i} \right) v^i + c_2 \frac{nT - v^i v^i}{2T^2} \frac{\partial T}{\partial t} + c_3 \frac{(n+2)Tv^i - v^j v^j v^i}{2T^2} \frac{\partial T}{\partial x^i} \right] G_c, \quad (5.7)$$

where $L^\dagger$ is the Fokker-Planck operator of the Klein-Kramers equation (A.1).

We are interested in the high-friction limit of this equation and therefore proceed as in appendix A.1 by rescaling the friction constant $\tilde{\gamma} = \epsilon \gamma$, which introduces a separation of timescales. The Feynman-Kac formula can then be reordered in powers of ϵ as

$$\frac{\partial G_c}{\partial t} - (L_0^\dagger + \epsilon^{-1} L_1^\dagger + \epsilon L_2^\dagger) G_c = 0, \quad (5.8)$$

where the different operators read

$$L_0^\dagger = -v^i \frac{\partial}{\partial x^i} - f^i \frac{\partial}{\partial v^i} - c_1 \left(\frac{f^i}{T} - \frac{1}{T} \frac{\partial T}{\partial x^i} \right) v^i - c_3 \frac{(n+2)Tv^i - v^j v^j v^i}{2T^2} \frac{\partial T}{\partial x^i}, \tag{5.9}$$

$$L_1^\dagger = \tilde{\gamma} \left[\frac{\partial}{\partial v^i} v^i + T \frac{\partial}{\partial v^i} \frac{\partial}{\partial v^i} \right], \tag{5.10}$$

$$L_2^\dagger = -c_2 \frac{nT - v^i v^i}{2T^2} \frac{\partial T}{\partial \tilde{t}}. \tag{5.11}$$

Here we have assumed that the temperature changes only on slow timescales $\tilde{t} = \epsilon t$. Note that $L_1^\dagger$ is the same operator as in equation (A.5) in the derivation of the high-friction limit of the Fokker-Planck equation. We now develop the solution as a power series in ϵ as $G_c = G^{(0)} + \epsilon G^{(1)} + \epsilon^2 G^{(2)}....$ To obtain the solutions of G_c at different orders in ϵ we proceed exactly as in appendix A.1 and therefore omit here the details of the calculations. The solution at order ϵ^{-1} becomes

$$G^{(0)} = q(x, t, \tilde{t}) \omega(v, x, \tilde{t}), \quad \omega(x, v, \tilde{t}) = \frac{e^{-\frac{v^i v^i}{2T}}}{(2\pi T)^{n/2}}. \tag{5.12}$$

At order ϵ^0 we obtain

$$G^{(1)} = \left[r - \frac{v^i}{\tilde{\gamma}} \frac{\partial q}{\partial x^i} - q(c_1 - 1) \left(\frac{f^i}{T} - \frac{1}{T} \frac{\partial T}{\partial x^i} \right) \frac{v^i}{\tilde{\gamma}} - \frac{q(c_3 - 1)}{6\tilde{\gamma}T^2} \frac{\partial T}{\partial x^i} ((n+2)Tv^i - v^i v^j v^j) \right] \omega, \tag{5.13}$$

where $r(x, \tilde{t})$ is independent of velocity. At order ϵ^1 after relaxation over fast timescales, equation (5.8) becomes

$$L_1^\dagger G^{(2)} = \frac{\partial G^{(0)}}{\partial \tilde{t}} - L_0^\dagger G^{(1)} - L_2^\dagger G^{(0)}. \tag{5.14}$$

This equation cannot be solved generally, but it is possible to obtain a solvability condition for it. Integrating over velocity and exploiting the orthogonality of the Hermite polynomials this condition reads

$$\frac{\partial q}{\partial t} + \frac{\partial}{\partial x^i} \left(\frac{f^i q}{\gamma} \right) - \frac{\partial}{\partial x^i} \left(\frac{1}{\gamma} \frac{\partial}{\partial x^i} (Tq) \right) - 2c_1 \frac{\partial}{\partial x^i} \left(\frac{f^i q}{\gamma} - \frac{\partial T}{\partial x^i} \frac{q}{\gamma} \right)$$

$$+ c_1 \frac{Tq}{\gamma} \frac{\partial}{\partial x^i} \left(\frac{f^i}{T} - \frac{1}{T} \frac{\partial T}{\partial x^i} \right) + c_1 q \left(\frac{f^i}{T} - \frac{1}{T} \frac{\partial T}{\partial x^i} \right) \left(T \frac{\partial \gamma^{-1}}{\partial x^i} + \frac{f^i}{\gamma} \right)$$

$$- c_1^2 q \frac{T}{\gamma} \left(\frac{f^i}{T} - \frac{1}{T} \frac{\partial T}{\partial x^i} \right) \left(\frac{f^i}{T} - \frac{1}{T} \frac{\partial T}{\partial x^i} \right) + \frac{qc_3(1 - c_3)}{6\gamma T} (n+2) \frac{\partial T}{\partial x^i} \frac{\partial T}{\partial x^i} = 0, \tag{5.15}$$

where we have already reverted to the original time and friction.
Since we are interested in the anomalous entropy production, we set $c_3 = 1$ and $c_1 = c_2 = 0$. Equation (5.15) then reduces to

$$\frac{\partial q}{\partial t} + \frac{\partial}{\partial x^i} \left(\frac{f^i q}{\gamma} \right) = \frac{\partial}{\partial x^i} \frac{1}{\gamma} \frac{\partial}{\partial x^i} (Tq), \tag{5.16}$$

which we identify as the Smoluchowski equation derived in equation (A.25). To lowest order we therefore obtain

$$\langle e^{-S_{anom}}\delta(X_t - x)\delta(V_t - v)\rangle \xrightarrow{\gamma \gg 1} G^{(0)} = \rho(x,t|x',t')\omega(x,v,t|x',v',t'), \qquad (5.17)$$

where we have replaced q by the solution for the probability density $\rho(x,t)$ of the Smoluchowski equation for overdamped trajectories. Averaging over all initial and final conditions, we are left with

$$\langle e^{-s_{anom}}\rangle = 1. \qquad (5.18)$$

This constitutes a general fluctuation relation for the s_{anom}.

To calculate the average anomalous entropy production rate we set $c_1 = c_2 = 0$ and use the property of the generating function

$$\frac{d}{dt}\langle s_{anom}\rangle = -\frac{d}{dt}\int dxdv \frac{\partial G_c}{\partial c_3}\Big|_{c_3=0} \xrightarrow{\gamma \gg 1} -\frac{d}{dt}\int dxdv \frac{\partial}{\partial c_3}(q\omega)\Big|_{c_3=0} = -\int dx \frac{\partial}{\partial c_3}\frac{\partial q}{\partial t}\Big|_{c_3=0}, \qquad (5.19)$$

where in the last step we integrated out the velocity and pulled the time derivative under the integral. We can now express $\frac{\partial q}{\partial t}$ via the solvability condition (5.15) as

$$\frac{d}{dt}\langle s_{anom}\rangle = \int dx \frac{\partial}{\partial c_3}\left[\frac{\partial}{\partial x^i}\left(\frac{f^i q}{\gamma} - \frac{1}{\gamma}\frac{\partial}{\partial x^i}(Tq)\right) + \frac{qc_3(1-c_3)}{6\gamma T}(n+2)\frac{\partial T}{\partial x^i}\frac{\partial T}{\partial x^i}\right]\Big|_{c_3=0}. \qquad (5.20)$$

The first part of the bracket goes to zero after integration and the second part after carrying out the partial derivative becomes

$$\frac{d}{dt}\langle s_{anom}\rangle = \int dx \frac{q(1-2c_3)}{6\gamma T}(n+2)\frac{\partial T}{\partial x^i}\frac{\partial T}{\partial x^i}\Big|_{c_3=0} = \int dx\rho \frac{(n+2)}{6\gamma T}\frac{\partial T}{\partial x^i}\frac{\partial T}{\partial x^i}, \qquad (5.21)$$

where we have used the fact that for $c_3 = 0$ q just becomes the overdamped probability distribution in space ρ. We finally conclude, that the introduction of a spatially inhomogeneous temperature profile, leads to an anomalous contribution to the entropy production, that stems from a higher order correction to the simple overdamped result. For high friction its average rate of production is given by equation (5.21), which is not taken into account if the the process is described by the overdamped Langevin equation (5.5).

5.2 Case Study - Quadratic Temperature Profile

To investigate the findings of the last section in a concrete model system, we consider stochastic particle motion in a one dimensional harmonic trap exposed to a quadratic temperature profile

$$V(x) = \frac{1}{2}\lambda x^2, \quad T(x) = T_0 + gx^2. \qquad (5.22)$$

This model is already so complicated that even in the overdamped limit the Fokker-Planck equation cannot be solved analytically. But if we assume that the system relaxes into a stationary solution, the probability flux must vanish, which in the overdamped limit means

$$J_x = -\frac{1}{\gamma}\left[T\frac{\partial \rho_s}{\partial x} + \left(\frac{\partial V}{\partial x} + \frac{\partial T}{\partial x}\right)\rho_s\right] = 0. \qquad (5.23)$$

Inserting equation (5.22) into equation (5.23) we obtain

$$\ln\left(\frac{\rho_s}{\rho_0}\right) = -\int dx \frac{(\lambda + 2g)x}{T_0 + gx^2} = -\left(1 + \frac{\lambda}{2g}\right)\ln(T_0 + x^2), \tag{5.24}$$

which after normalization becomes

$$\rho_s = \frac{\Gamma\left(1 + \lambda/(2g)\right)}{\Gamma\left(1/2 + \lambda/(2g)\right)}\sqrt{\frac{g}{\pi T_0}}\left(\frac{T_0}{T_0 + gx^2}\right)^{1+\lambda/(2g)}, \tag{5.25}$$

where Γ denotes the Gamma-function. The stationary distribution is of power law type which gives us conditions for the finiteness of the nth moments

$$\langle x^n \rangle = \begin{cases} 0 & n \text{ odd} \\ \frac{\Gamma[(1+n)/2]\Gamma[(1-n)/2+\lambda/(2g)]}{\sqrt{\pi}\Gamma[1/2+\lambda/(2g)]}\left(\frac{T_0}{g}\right)^{n/2}, & n \text{ even}, \lambda > g(n-1) \\ \infty & n \text{ even } \lambda < g(n-1). \end{cases} \tag{5.26}$$

This basically means that with respect to the nth moment the potential is not confining for $\lambda < g(n - 1)$. The average regular entropy production is of the form $\int \langle g(x)\dot{x}\rangle d\tau$ (see equation (5.3)) and it can be shown generally[21] that

$$\langle g(x)\dot{x} \rangle = \int dx g(x) J_x(x, t). \tag{5.27}$$

After relaxation to the stationary distribution, the probability flux J_x vanishes and so does the regular entropy production. In that case the only remaining part in the entropy production is the anomalous contribution from equation (5.21)

$$\frac{d}{dt}\langle s_{env}\rangle = \frac{d}{dt}\langle s_{anom}\rangle = \left\langle \frac{1}{2\gamma T}\frac{\partial T}{\partial x^i}\frac{\partial T}{\partial x^i}\right\rangle = \int \frac{2g^2x^2}{\gamma(T_0 + gx^2)}\rho_s dx = \frac{2g^2}{\gamma(2g + \lambda)}. \tag{5.28}$$

To verify the findings, we simulated the underdamped dynamics of the system and started to measure the entropy production in the environment after the distribution had relaxed to the stationary distribution. In figure 5.1(a) the analytical and numerical stationary distribution is depicted for a parameter set of $T_0 = 1, g = 1, \lambda = 5$ and two different damping constants $\gamma = 100, \gamma = 10$. We see that the stationary distributions agree very well. This is to be expected, because for a static potential and a static temperature, the only timescale in the system is the relaxation time of the velocity $\tau_{rel} = 1/\gamma$. This means that after relaxation, the overdamped and underdamped stationary distributions will be the same, because the overdamped assumption $t\gamma \gg 1$ is for long enough times t fulfilled for arbitrary γ.

Figure 5.1(b) shows the rate of entropy production in the environment after the system has relaxed to the stationary state. We see that the analytical prediction and the numerical result agree very well. In addition we also simulated only the overdamped Langevin dynamics (equation (5.5)). We indeed did not see any entropy production after the system had relaxed to the stationary distribution. This indicates that even when the overdamped trajectories reproduce the ensemble distribution in position, it is necessary to simulate the full dynamics when one is interested in entropy production in the environment in the presence of temperature gradients.

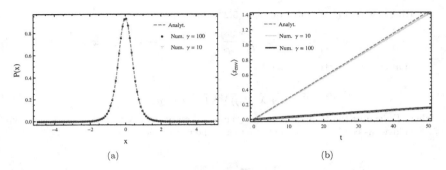

(a)   (b)

Figure 5.1: (a) Comparison of analytical overdamped stationary solution (equation (5.25)) and numerical solution for the full dynamics after relaxation in a thermal environment with a quadratic temperature gradient. (b) Comparison of the analytical prediction (equation (5.28)) and numerical calculation of the entropy production in the environment. Parameter set: $T_0 = 1, g = 1, \lambda = 5$.

5.3   Entropy Production in Optical Trap Setup

The anomalous entropy production has so far not been measured experimentally. One of the reasons is that the generation and control of temperature gradients on small scales is very difficult to achieve. As we have seen in chapter 3.2.1 it is possible to control an effective temperature by means of optical damping via a control laser beam. Therefore the question arose, whether optical damping could also be used to create an effective temperature gradient in the trap setup described in chapter 3.2 and if subsequently an anomalous entropy production could be measured. We therefore numerically studied the complete three-dimensional set of Langevin equations (3.3). In the harmonic approximation and with dimensionless units, they take the form

$$dv_x = [-\gamma_{th}v_x - \lambda_r x]\, dt + \sqrt{2\gamma_{th}T}dW_x,$$
$$dv_y = [-\gamma_{th}v_y - \lambda_r y]\, dt + \sqrt{2\gamma_{th}T}dW_y,$$
$$dv_z = [-\gamma_{eff}v_z - \lambda_z z]\, dt + \sqrt{2\gamma_{eff}T_{eff}}dW_z,$$

(5.29)

$$\text{with}\quad \gamma_{eff} = \gamma_{th}(1 + fe^{-(x^2+y^2)}),\quad T_{eff} = \frac{T}{(1 + fe^{-(x^2+y^2)})},\quad \gamma_{opt} = f\gamma_{th}.$$

Here we have defined an effective damping constant γ_{eff} in z-direction, by combining the optical and thermal damping. As a consequence, we also obtain an effective temperature T_{eff}, where $T\gamma_{th} = T_{eff}\gamma_{eff}$. Since T_{eff} depends on x and y, an anomalous entropy production seemed plausible.

In figure 5.2 the numerically calculated mean entropy production in the environment is shown over time, where all parameters have been adjusted to the experimental trap parameters (see equation (3.5)). The initial conditions were sampled from the equilibrium distribution when the optical damping was switched off. We see that after a relaxation

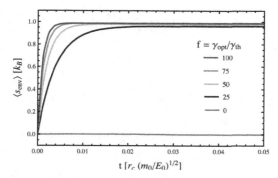

Figure 5.2: Mean entropy production in the optical trap setup described in chapter 3.2. The entropy production vanishes after relaxation to the stationary distribution. Units are given in $E_0 = 300K \, k_B, r_c = 40 \cdot 10^{-6}m, m_0 = 10^{-17}kg$.

process towards the stationary distribution, the entropy production vanishes, independent of the strength f of the optical damping. The trap setup, as it is, will therefore not be able to measure the anomalous entropy production. The most probable reason for this is the following. In the derivation of the anomalous entropy production at the beginning of this chapter the temperature was supposed to be a scalar field. This means that at a given point in space the fluctuating force due to the interaction with the temperature bath is isotropic in the different spatial directions. In equation (5.29) on the other hand, the effective temperature $T_{\text{eff}}(x, y)$ only enters the z component. Therefore, the particle in the trap experiences stronger fluctuations (and also stronger damping) in the z-direction than in the other directions. This was not included in the derivation of the anomalous entropy production. Furthermore, the spatial dependence in the effective temperature $T_{\text{eff}}(x, y)$ is only on the x and y components. Consequently, there is no spatial variation in temperature along the z-direction itself in which the thermal fluctuations enter. Besides, the motion in radial direction decouples from the z-direction due to the harmonic approximation of the potential, so that no feedback from the motion in z-direction enters back into the radial one. However, for a rigorous explanation of why we do not see anomalous entropy production in our simulation one would have to generalize the derivation of section 5.1 to the case of non-isotropic fluctuations and damping constants, which was out of scope within the limits of this thesis.

For a real temperature gradient along the cavity axis, on the other hand, we expect to observe anomalous entropy production. For a quadratic temperature profile of the form of Eq. (5.22) along the z-direction we show the rate of entropy production in figure 5.3 over the gradient strength g.

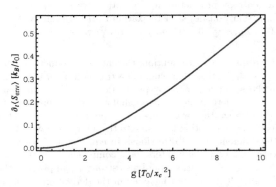

Figure 5.3: Rate of entropy production for a quadratic temperature profile along the cavity axis is depicted over the strength g of the temperature gradient (see equation (5.22)). Units are given in $t_0 = x_c(k_B T_0/m_0)^{-1/2}$, $m_0 = 10^{-17} kg$, $T_0 = 295K$, $x_c = 40 \cdot 10^{-6} m$.

6

Conclusions and Outlook

In this thesis we studied a stochastic heat engine in the overdamped and underdamped limit. We extended the theory of optimal overdamped driving protocols for a Brownian particle in a harmonic trapping potential to the case of an optical cavity trap setup in which the temperature difference between the two heat baths is created by optical damping. Furthermore we showed that in the underdamped limit the optimal protocols derived in the overdamped approximation fail. This is particularly true for the experimentally accessible parameter range in the optical cavity trap setup described in section 3.2.

We therefore generalized the overdamped optimization theory to the underdamped regime. For that purpose we derived an approximate Fokker-Planck equation governing the probability distribution in the underdamped limit and applied it to the case of a harmonic trapping potential. The full optimization problem is analytically and numerically very hard to tackle. As a consequence, we systematically approximated the genuine optimal protocol from linear segments. Starting with a simple linear driving protocol we were able to obtain analytical solutions for the probability distribution and numerically optimized the free protocol parameters to obtain a maximal power output of the heat engine. Splitting the linear protocol into piecewise linear protocols and optimizing the additional free parameters we showed that the power output could be further improved. We therefore expect to come arbitrarily close to the genuine optimal protocol in the underdamped regime by applying this technique iteratively. At maximum power, the optimized simple linear and the piecewise-linear protocols were qualitatively very similar. We therefore do not expect a qualitative difference of the protocol by subdividing it any further.

A possible continuation of this work would be to study specific nonlinear protocols and optimize the associated free protocol parameters. Here, the same iterative splitting technique could be employed. Furthermore, combinations of different functional forms for the different protocol segments are conceivable, especially for the coupling to the hot and cold bath, respectively.

In the last chapter we reviewed the recently discovered anomalous entropy production [24] in the presence of temperature gradients and found analytical expressions for the case of a quadratic temperature profile that we also verified numerically. A natural extension of this work would consequently be to study optimal protocols of a stochastic heat engine in the presence of temperature gradients. In this context it has been shown that already small temperature gradients severely affect the efficiency in the slow-driving quasi-static limit [49]. For this reason we expect that the integration of temperature gradients in the analysis of the stochastic heat engine will have a strong qualitative effect.

Throughout the investigations presented in this thesis, we were in close communication with the experimental group around Nikolai Kiesel and Markus Aspelmeyer at the University of Vienna. They are running a sophisticated optical trap setup [36] and expressed a strong interest in realizing a stochastic heat engine in their experiment. Until the due date of this thesis no experimental data was collected, but we expect a quantitative comparison of our findings to experimental data to be possible in the near future.

A

Appendix

A.1 Derivation of the Smoluchowski Equation

In chapter 5 we are interested in systems which have a position-dependent temperature profile. Therefore, the subsequent derivation is carried out as general as possible and friction and temperature are possibly position-dependent, $\gamma = \gamma(x), T = T(x)$ without indicating this dependence all the time.

Our starting point is the Klein-Kramers equation (1.22) in n dimensions

$$\frac{\partial P(x,t)}{\partial t} = L_{FP}^\dagger P(x,t), \quad L_{FP}^\dagger = -v^i \frac{\partial}{\partial x^i} + \frac{\partial}{\partial v^i} \left(-f^i + \gamma v^i\right) + \gamma T \frac{\partial^2}{\partial v^i \partial v^i}, \tag{A.1}$$

where we use the Einstein summation convention for repeated indices. $f^i = -\partial V/\partial x^i$ is the force exerted by an external potential, x^i are the components of the n-dimensional vector $\mathbf{x}$ and we have rescaled to dimensionless units. To reobtain the actual values, one has to multiply by the following units

$$[v] = \sqrt{\frac{k_B T_0}{m}}, \quad [x] = [x_c], \quad [t] = x_c/\sqrt{\frac{k_B T_0}{m}}, \quad [\gamma] = \frac{\sqrt{k_B T_0 m}}{x_c}, \quad [T] = T_0, \quad [f] = \frac{k_B T_0}{x_c}, \tag{A.2}$$

where T_0 is a reference temperature, x_c is a typical length scale of the system, m is the particle mass and k_B is the Boltzmann constant.

Since we assume the damping to be high, we rescale the friction coefficient by a small bookkeeping parameter ϵ, i.e. $\gamma = \tilde{\gamma}/\epsilon$. For small ϵ this introduces a timescale separation, where we distinguish between three different timescales treated as independent variables. The fast timescale is given by $\theta = \epsilon^{-1}t$ and we can associate it to frictional relaxation. The intermediate timescale is given by t, and the slow timescale is defined by $\tilde{t} = \epsilon t$ and determines the overdamped dynamics.

The transition probability which in this context may also be called the propagator is assumed to be a function of the different timescales and in the limit of $\epsilon \to 0$ can be developed in a power series of ϵ

$$\mathcal{P} = p^{(0)} + \epsilon p^{(1)} + \epsilon^2 p^{(2)} .... \tag{A.3}$$

The Fokker-Planck operator can then be split into parts operating on different timescales, so that the Fokker-Planck equation reads

$$\left(\frac{\partial}{\partial t} - L_{FP}^\dagger\right) p = \left(\frac{\partial}{\partial t} - L_0^\dagger - \epsilon^{-1} L_1^\dagger\right) p = 0, \tag{A.4}$$

where the operators are given by

$$L_0^\dagger = -v^i \frac{\partial}{\partial x^i} - f^i \frac{\partial}{\partial v^i}, \quad L_1^\dagger = \tilde{\gamma}\left[\frac{\partial}{\partial v^i}v^i + T\frac{\partial}{\partial v^i}\frac{\partial}{\partial v^i}\right]. \tag{A.5}$$

To find a solution to $p^{(0)}$, we plug the expansion (A.3) into equation (A.4) and only consider terms at order ϵ^{-1} which leads to

$$\left(\frac{\partial}{\partial\theta} - L_1^\dagger\right)p^{(0)} = 0, \tag{A.6}$$

and thus is the equation governing the dynamics at fast timescales. We can write down the formal solution of equation (A.6)

$$p^{(0)} = e^{L_1^\dagger\theta}p_{\theta=\theta_0}^{(0)}. \tag{A.7}$$

Note that since $L_1^\dagger$ is not explicitly time-dependent we do not have to use the time-ordering operator. We now want to calculate the eigenfunctions of the operator $L_1^\dagger$. Realizing that

$$L_1^\dagger e^{-\frac{v^i v^i}{2T}}p = \tilde{\gamma}e^{-\frac{v^i v^i}{2T}}\left[-v^i\frac{\partial}{\partial v^i} + T\frac{\partial}{\partial v^i}\frac{\partial}{\partial v^i}\right]p, \tag{A.8}$$

we can identify the term in brackets on the right hand side as the differential operator of Hermite's differential equation (see appendix A.2). Thus the eigenfunctions $\psi_{\lambda_1,\dots,\lambda_n}$ of $L_1^\dagger$ are products of the weight function $\omega(x,v)$ and the Hermite polynomials

$$L_1^\dagger\psi_{\lambda_1,\dots,\lambda_n} = -\tilde{\gamma}\left(\sum_{i=1}^n \lambda_i\right)\psi_{\lambda_1,\dots,\lambda_n},$$

$$\psi_{\lambda_1,\dots,\lambda_n} = \omega(x,v)\prod_{i=1}^n He_{\lambda_i}(v^i/\sqrt{T}), \quad \omega(x,v) = \frac{e^{-\frac{v^i v^i}{2T}}}{(2\pi T)^{n/2}}, \tag{A.9}$$

where $He(x)$ are the Hermite polynomials and n is the dimensionality of the system. These eigenfunctions form a basis in the space of probability distributions in which we can decompose $p_{\theta=\theta_0}^{(0)}$ from equation (A.7), leading to

$$p^{(0)} = \sum_{\lambda_1,\dots,\lambda_n} c_{\lambda_1,\dots,\lambda_n} e^{-\tilde{\gamma}\left(\sum_{i=1}^n \lambda_i\right)\theta}\psi_{\lambda_1,\dots,\lambda_n}. \tag{A.10}$$

The spectrum of $L_1^\dagger$ is negative and with $\theta \to \infty$ the solution therefore relaxes on fast timescales exponentially fast to the zeroth eigenfunction

$$p^{(0)}(x,v,t,\tilde{t}) = \rho(x,t,\tilde{t})\omega(x,v), \quad c_{0,\dots,0} = \rho(x,t,\tilde{t}). \tag{A.11}$$

Note that $c_{0,\dots,0} = \rho(x,t,\tilde{t})$ has to be independent of v, because of equation (A.8). That means that the velocity dependence of $p^{(0)}$ comes in only through the weight function $\omega(x,v)$ and $p^{(0)}$ is therefore just the Maxwell-Boltzmann distribution in velocity space with the local temperature $T(x)$ times the marginal probability density in space at lowest order.

Next, we look at intermediate timescales where we take only terms of order ϵ^0 into account and equations (A.4) and (A.3) give

$$\left(\frac{\partial}{\partial\theta} - L_1^\dagger\right) p^{(1)} = -\left(\frac{\partial}{\partial t} - L_0^{(0)}\right). \tag{A.12}$$

After relaxation on fast timescales $p^{(1)}$ does not depend on θ and we can neglect the $\frac{\partial}{\partial\theta}$ term leading to

$$L_1^\dagger p^{(1)} = \left(\frac{\partial}{\partial t} - L_0^\dagger\right) p^{(0)}. \tag{A.13}$$

Integrating both sides over v, the left side vanishes because the distribution has to go to zero at the boundary

$$\int_{-\infty}^\infty dv\,\tilde\gamma\left[\frac{\partial}{\partial v}v + T\frac{\partial}{\partial v}\frac{\partial}{\partial v}\right] p^{(1)} = \tilde\gamma\left[\frac{\partial}{\partial v}v + T\frac{\partial}{\partial v^i}\frac{\partial}{\partial v}\right] p^{(1)}\bigg|_{-\infty}^\infty = 0. \tag{A.14}$$

The right hand side of equation (A.13) becomes

$$\int_{-\infty}^\infty dv\frac{\partial}{\partial t}\rho(x,t,\tilde t)\omega(x,v) + \int_{-\infty}^\infty \left(dv v^i\frac{\partial}{\partial x^i} + f^i\frac{\partial}{\partial v^i}\right)\rho(x,t,\tilde t)\omega(x,v). \tag{A.15}$$

The last term in brackets vanishes for the same reason as before, namely that the distribution is bounded. The first term in brackets vanishes, because the v dependence comes only through the term $v\omega(x,v)$, which is an odd function in v. Therefore, we arrive at

$$\frac{\partial\rho(x,t,\tilde t)}{\partial t} = 0, \tag{A.16}$$

which is a solvability condition and implies that $\rho(x,\tilde t)$ only depends on slow timescales $\tilde t$. Using equation (A.16) in equation (A.13) we have

$$L_1^\dagger p^{(1)} = -L_0^\dagger p^{(0)} = \left(v^i\frac{\partial}{\partial x^i} + f^i\frac{\partial}{\partial v^i}\right)\rho\omega = \omega v^i\frac{\partial\rho}{\partial x^i} + \rho v^i\frac{\partial\omega}{\partial x^i} + \rho f^i\frac{\omega}{\partial v^i} \tag{A.17}$$

$$= \omega v^i\frac{\partial\rho}{\partial x^i} + \rho\omega v^i\frac{(v^j v^j - nT)}{2T^2}\frac{\partial T}{\partial x^i} - \rho\frac{f^i v^i}{T}\omega. \tag{A.18}$$

To reconstruct $p^{(1)}$, we notice that

$$L_1^\dagger(v^i\omega) = -\tilde\gamma v^i\omega, \quad L_1^\dagger\left[((n+2)Tv^i - v^i v^j v^j)\omega\right] = -3\tilde\gamma((n+2)Tv^i - v^i v^j v^j)\omega. \tag{A.19}$$

Identifying terms in equation (A.18) with terms in equation (A.19) we finally obtain

$$p^{(1)} = r\omega - \frac{v^i}{\tilde\gamma}\frac{\partial\rho}{\partial x^i}\omega + \frac{v^i}{\tilde\gamma T}\left(f^i - \frac{\partial T}{\partial x^i}\right)\rho\omega + \frac{((n+2)T - v^j v^j)v^i}{6\tilde\gamma T^2}\frac{\partial T}{\partial x^i}\rho\omega, \tag{A.20}$$

where $r(x,\tilde t)$ is a possible contribution from the null-space of $L_1^\dagger$, i.e. $L_1^\dagger(r\omega) = 0$ which immediately follows from equation (A.8).

We now come to slow time-scales which determine the overdamped limit. At order ϵ^1 we have from equation (A.4) after relaxation over the fast variables,

$$L_1^\dagger p^{(2)} = -L_0^\dagger p^{(1)} + \frac{\partial}{\partial \tilde{t}} p^{(0)}. \tag{A.21}$$

To obtain the solvability condition we again integrate both sides over v. The left side vanishes for the same reason as before (see equation (A.14)) leaving us with

$$\int_{-\infty}^{\infty} dv \left(-v^i \frac{\partial}{\partial x^i} - \underset{\underbrace{\quad}}{f^i \frac{\partial}{\partial v^i}} \right) p^{(1)} = -\int_{-\infty}^{\infty} dv v^i \frac{\partial}{\partial x^i} p^{(1)} = \int_{-\infty}^{\infty} dv \frac{\partial}{\partial \tilde{t}} p^{(0)}, \tag{A.22}$$

where the underlined part goes to zero after integrating by parts. Inserting $p^{(1)}$ from equation (A.20) in equation (A.22) we obtain

$$-\int_{-\infty}^{\infty} dv v^i \frac{\partial}{\partial x^i} \left[r\omega - \frac{v^i}{\tilde{\gamma}} \frac{\partial \rho}{\partial x^i} \omega + \frac{v^i}{\tilde{\gamma}T} \left(f^i - \frac{\partial T}{\partial x^i} \right) \rho\omega + \frac{((n+2)T - v^jv^j)v^i}{6\tilde{\gamma}T^2} \frac{\partial T}{\partial x^i} \rho\omega \right]$$
$$= \int_{-\infty}^{\infty} dv \frac{\partial}{\partial \tilde{t}} p^{(0)}. \tag{A.23}$$

The first part in brackets vanishes after integration, because $vr\omega$ is odd in v. The last part in brackets is a product of different Hermite polynomials times the weight function $\omega(x, v)$ and vanishes due to the orthogonality of the Hermite polynomials (see appendix A.2). After integrating the remaining parts over velocities, the solvability condition becomes

$$\frac{\partial \rho}{\partial \tilde{t}} = -\frac{\partial}{\partial x^i} \left(\frac{f^i}{\tilde{\gamma}} \rho \right) + \frac{\partial}{\partial x^i} \frac{1}{\tilde{\gamma}} \frac{\partial}{\partial x^i} T\rho, \tag{A.24}$$

and reverting to the original time-scales $t = \epsilon^{-1}\tilde{t}$ and friction $\gamma = \epsilon^{-1}\tilde{\gamma}$ we finally obtain

$$\frac{\partial \rho}{\partial t} = -\frac{\partial}{\partial x^i} \left(\frac{f^i}{\gamma} \rho \right) + \frac{\partial}{\partial x^i} \frac{1}{\gamma} \frac{\partial}{\partial x^i} T\rho. \tag{A.25}$$

For position-independent temperature this is exactly the Smoluchowski equation (1.44). Note that if the temperature depends on position, the probability flux

$$j^i = -\frac{T}{\gamma} \frac{\partial \rho}{\partial x^i} + \frac{1}{\gamma} \left(f^i - \frac{\partial T}{\partial x^i} \right) \rho \tag{A.26}$$

has an extra term in which enters the gradient of the temperature.

A.2 Hermite Polynomials

The Hermite polynomials are defined as

$$He_\lambda(x) = (-1)^\lambda e^{x^2/2} \frac{d^\lambda}{dx^\lambda} e^{-x^2/2} \tag{A.27}$$

and solve the differential equations

$$- u''(x) + 2xu'(x) = 2\lambda u(x) \quad \lambda = 0, 1, ..., n. \tag{A.28}$$

This means that He_λ are the eigenfunctions of the differential operator

$$\hat{L} = -\frac{1}{2} \frac{\partial^2}{\partial x^2} + x \frac{\partial}{\partial x} \tag{A.29}$$

with eigenvalue λ. The Hermite polynomials are orthogonal with respect to the measure

$$\int_{-\infty}^{\infty} He_m(x) He_n(x) \omega(x) dx = \delta_{mn}, \quad \omega(x) = \frac{1}{\sqrt{2\pi}} e^{-x^2/2} \tag{A.30}$$

and form an orthogonal basis of the Hilbert space. The first Hermite Polynomials read

$$He_0(x) = 1, \quad He_1(x) = x, \quad He_2(x) = x^2 - 1, \quad He_3(x) = x^3 - 3x, ... \tag{A.31}$$

For further reading see [61].

A.3 Simulation Method

Langevin dynamics was first introduced in molecular simulations to calculate the properties of mesoscopic systems [62], where a dissipative force and a noise term were added to the Hamilton equations to model a bath of lighter particles. Many simulation algorithms have since then been proposed, in this work we use the one introduced by Bussi and coworkers [63] and elaborated by Sivak et. al.[64]. All programs were written in C++ and in the following we briefly outline the used integration scheme.

A.3.1 Integration Scheme

The integration scheme that we are seeking should allow the construction of single particle trajectories according to the Langevin equation (1.1). The probability distribution related to the ensemble of trajectories is on the other hand supposed to obey the according Fokker-Planck equation. For the derivation of the integration scheme we therefore start with the formal solution of the latter

$$P(v, x, t + \Delta t) = e^{-\Delta t \hat{L}_{FP}} P(v, x, t), \tag{A.32}$$

with

$$\hat{L}_{FP} = -\frac{1}{m} \frac{\partial V(x)}{\partial x} \frac{\partial}{\partial v} + v \frac{\partial}{\partial x} - \frac{\gamma}{m} \left(\frac{\partial}{\partial v} v + kT \frac{\partial}{\partial v} \frac{\partial}{\partial v} \right). \tag{A.33}$$

As first realized by Tuckerman et al. [65] the Trotter splitting formula [66] allows an approximated propagator to be constructed as

$$e^{\Delta \hat{L}} \approx \prod_{j=M}^{1} e^{-\frac{\Delta t}{2} \hat{L}_j} \prod_{k=1}^{M} e^{-\frac{\Delta t}{2} \hat{L}_k}. \tag{A.34}$$

Here M is the number of stages in the integrator and $\sum_j \hat{L}_j = \hat{L}$. The splitting is approximate since in general the $\hat{L}_j$'s do not commute. Therefore the order in which the stages are applied is relevant. Note that here a symmetric construction of the propagator is chosen which will lead to a symmetric integrator. The main idea about the splitting is, that the different stages $e^{-\frac{\Delta t}{2}\hat{L}_j}$ are chosen so that they can be integrated analytically and the Trotter splitting remains the only source of errors. For the Fokker-Planck operator, the natural choice of decomposition is

$$\hat{L}_{FP} = \hat{L}_v + \hat{L}_x + \hat{L}_\gamma, \tag{A.35}$$

with

$$\hat{L}_v = -\frac{1}{m}\frac{\partial V(x)}{\partial x}\frac{\partial}{\partial v}, \quad \hat{L}_x = v\frac{\partial}{\partial x}, \quad \hat{L}_\gamma = -\frac{\gamma}{m}\left(\frac{\partial}{\partial v}v + kT\frac{\partial}{\partial v}\frac{\partial}{\partial v}\right). \tag{A.36}$$

We notice that the operator $e^{-\frac{\Delta t}{2}\hat{L}_\gamma}$ leaves the stationary distribution

$$P_S(v,x)dvdx \propto e^{-\beta(\frac{mv^2}{2}+\beta V(x))}dvdx \tag{A.37}$$

invariant

$$e^{-\frac{\Delta t}{2}\hat{L}_\gamma}P_S = P_S. \tag{A.38}$$

This is due to the fact that $\hat{L}_\gamma$ introduces the combined effects of noise and friction under which the stationary distribution should not change. The splitting that we are going to use is the same used to obtain the velocity Verlet algorithm, namely

$$e^{-\Delta t\hat{L}} \approx e^{-\frac{\Delta t}{2}\hat{L}_\gamma}e^{-\frac{\Delta t}{2}\hat{L}_v}e^{-\Delta t\hat{L}_x}e^{-\frac{\Delta t}{2}\hat{L}_v}e^{-\frac{\Delta t}{2}\hat{L}_\gamma}. \tag{A.39}$$

The integration scheme for single particle trajectories which results from equation (A.39) then reads

$$v(n+\tfrac{1}{4}) = \sqrt{a}v(n) + \sqrt{\frac{1-a}{\beta m}}\mathcal{N}(n) \tag{A.40a}$$

$$v(n+\tfrac{1}{2}) = v(n+\tfrac{1}{4}) + \frac{\Delta t}{2}\frac{f(n)}{m} \tag{A.40b}$$

$$r(n+\tfrac{1}{2}) = r(n) + \frac{\Delta t}{2}v(n+\tfrac{1}{2}) \tag{A.40c}$$

$$\mathcal{H}(n) \rightarrow \mathcal{H}(n+1) \tag{A.40d}$$

$$r(n+1) = r(n+\tfrac{1}{2}) + \frac{\Delta t}{2}v(n+\tfrac{1}{2}) \tag{A.40e}$$

$$v(n+\tfrac{3}{4}) = v(n+\tfrac{1}{2}) + \frac{\Delta}{2}\frac{f(n+1)}{m} \tag{A.40f}$$

$$v(n+1) = \sqrt{a}v(n+\tfrac{3}{4}) + \sqrt{\frac{1-a}{\beta m}}\mathcal{N}'(n+1), \tag{A.40g}$$

where $f(n)$ is the force due to an external potential, $a = \exp(-\Delta t\gamma)$, and $\mathcal{N}$ and $\mathcal{N}'$ are independent, normally distributed random variables with zero mean and unit variance created by a Box-Muller algorithm [48]. Equations (A.40a) and (A.40g) follow from the fact that the propagator $e^{-\frac{\Delta t}{2}\hat{L}_\gamma}$ in equation (A.39), can be integrated analytically and corresponds to an Ornstein-Uhlenbeck process in the velocity variable. The solution of this process [25] is a Gaussian

$$P(v,t|v',t') = \sqrt{\frac{\beta m}{2\pi}(1 - e^{-2\gamma(t-t')})} \exp\left[-\frac{\beta m(v - e^{-\gamma(t-t')}v')^2}{2(1 - e^{-2\gamma(t-t')})}\right]. \tag{A.41}$$

Making the identification $\Delta t/2 = t - t'$, we see that this is exactly the propagation enforced by equations (A.40a) and (A.40g).

Equations (A.40b)–(A.40f) correspond to the usual velocity Verlet algorithm where the Hamiltonian update is made explicit. This is for example relevant if we have a time-dependent potential. The integration steps follow immediately from the contribution of $\hat{L}_v$ and $\hat{L}_x$ in equation (A.39). The exponentials have been expanded up to first order in Δt since unlike as for L_γ the propagators cannot be integrated analytically. The velocity Verlet algorithm integration scheme is symplectic, this means that the Jacobian of the transformation from old to new positions and velocities is unity, and therefore the phase space volume is conserved [67]. Although a symplectic integrator does not conserve the energy of the system Hamiltonian, it does conserve the energy of a shadow Hamiltonian, which is close to the desired one for small time steps [68]. Our integration scheme is then time-symmetric and cleanly separates the stochastic and deterministic parts of the dynamics, where the deterministic parts are symplectic and the stochastic parts are detailed balanced. An advantage of this algorithm is that the work, shadow work and heat can be examined separately which is very convenient especially for the error analysis of the simulation [64].

A.4 Jumps in the Protocol at the Bath Change

So far we did not consider possible jumps in the protocol at the temperature bath change as they occur for optimal protocols in the overdamped limit (see section 2.2). Besides, it has been shown by Gomez-Marin and Seifert [69], that optimal protocols, which minimize the work needed to change the stiffness of a harmonic trap from λ_i to λ_f at a single bath temperature, exhibit jumps and even delta peaks. It therefore seems necessary to include possible discontinuous protocols in the optimization procedure. However, it is important to note that in Ref. [69] the delta and step functions occurred only at the beginning and at the end of the investigated protocol and irreversible entropy production occurring directly afterwards was not taken into account. Besides, they prepared the system in equilibrium. On the contrary, we are looking at a cyclic, in principle infinite process that is out of equilibrium at all times. Furthermore, in our case the aim is not to minimize the work needed to change the trap stiffness, but to maximize the work output gained through the alternated coupling to two different temperature baths. The setting, although at first sight seeming similar, is therefore quite different and their findings do not generally imply that jumps and delta peaks in the protocol improve the optimization process.

Concerning the occurrence of jumps in optimal protocols for the heat engine in the overdamped limit, we point out the importance of timescales on which the relevant processes take place. In the overdamped limit one is only looking at timescales where the velocities are always assumed to have already relaxed to the Maxwell distribution corresponding to the current temperature bath. Jumps in the protocol occurring in this limit hence correspond only to timescales much greater than the one on which velocity relaxes and the protocol might actually be smooth when one considers smaller timescales. In the underdamped regime we are looking exactly at scales on which velocity relaxation takes place. Therefore one has to pose the question of whether allowing jumps in this limit is physically still reasonable.

From an experimentalist's point of view, jumps in the protocol are not feasible, since the maximal rate of change of the trap stiffness is limited by the decay rate of the intensity of the laser[1]. Still one might argue that experimental techniques will advance in the future. Experimental protocols, however, are always continuous and can at best approximate discontinuities.

If we nevertheless decide to investigate jumps in the protocol from a theoretical point of view, we encounter a problem intrinsic to the approximation we made in order to derive the equations of motion (4.23). Discontinuities in the protocol translate into jumps in either the variance in position or in velocity. If we consider a discontinuity in the protocol at time t, i.e. $\Delta\lambda = \lambda(t^+) - \lambda(t^-) = \lambda^+ - \lambda^-$ the variances are related by (see equation (4.24))

$$w_v^+ = \lambda^+ w_x^+, \quad w_v^- = \lambda^- w_x^-. \tag{A.42}$$

It follows that only either w_v or w_x can be continuous, but not both at the same time. Clearly w_x should be continuous since position follows velocity. This can also be motivated by considering, that even in the single particle approach of the Langevin equation the actual trajectories are continuous but not differentiable.

It follows that if we choose w_x continuous, a jump in the protocol leads to a jump in w_v. Note that in our approximation w_v is not only the mean square velocity, but also the mean energy of the system. A sudden jump in the potential therefore naturally translates into a jump in energy. The problem is, that in deriving the Fokker-Planck equation (4.21) and everything subsequently built on it, we assumed the energy to be a slowly changing variable. For sudden jumps in the potential our approximation might therefore not be valid anymore. In an attempt to evaluate whether or not it is reasonable to investigate jumps within this framework, we allowed jumps in the protocol from equation (4.29) at the bath change, solved the equations of motion from equation (4.23) and enforced continuity on w_x to calculate the integration constants. We then compared the approximate solution to the numerical exact solution for different parameters. Exemplarily figure A.1 shows the comparison of the numerical exact and approximate analytical solution for the variances for an explicit parameter set of $T_h = 4$, $T_c = 1$ and $\gamma_{th} = 0.001$. The protocol of the trap stiffness $\lambda(t)$ is depicted in figure A.1(c). We see that indeed the approximation differs considerably from the exact numerical solution. This translates into unreliable predictions of power and efficiency and an optimization with jumps in the protocol as free parameters seems unreasonable. It is therefore not possible to investigate jumps in the protocol in this limit.

---

[1]This is at least true for the trap setups considered in chapter 3.

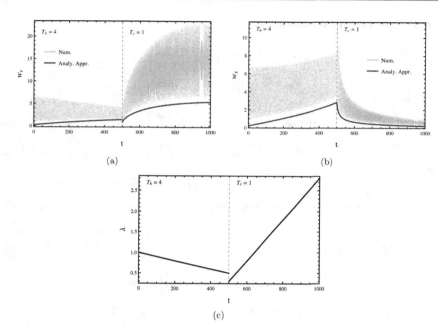

Figure A.1: Comparision of analytical approximation and numerical exact solution for a protocol with discontinuities. (a) Mean square velocity, (b) mean square position and (c) protocol of trap stiffness $\lambda(t)$ over time. Parameter set $T_h = 4, T_c = 1, \gamma_{th} = 0.001$. The analytical approximate solution fails in the presence of jumps, because the assumption of slowly varying energy is not fulfilled.

A.5   Integration Constants

$$C_{h1} = \left[-\frac{a}{2} e^{\frac{t_{tot}\left(\left(\frac{a\gamma_c}{2}+\gamma_c+\frac{\gamma_h}{4}\right)b^2-\left(\left(\gamma_c+\frac{\gamma_h}{4}\right)c_c+\left(\frac{a}{2}+1\right)c_h\gamma_c\right)b+c_cc_h\gamma_c\right)}{(a+1)(b-c_c)(b-c_h)}} F\left(\frac{\sqrt{-\frac{bt_{tot}\gamma_h}{a+1}}}{\sqrt{2}(c_h-b)}\right) \sqrt{b\,(b-c_c)\,c_c\,(1-c_h)\,c_h(1-c_c)\gamma_c}\,T_h\gamma_h t_{tot}^2 \right.$$

$$+ \frac{1}{2}ae^{\frac{t_{tot}\left(\left(\frac{a\gamma_c}{2}+\gamma_c\right)b^2-\left(\left(\frac{a\gamma_c}{2}+\gamma_c-\frac{\gamma_h}{2}\right)c_h+c_c\gamma_c\right)b+\left(\gamma_c-\frac{\gamma_h}{2}\right)c_cc_h\right)}{(a+1)(b-c_c)(b-c_h)}} F\left(\frac{\sqrt{-\frac{c_ht_{tot}\gamma_h}{a+1}}}{\sqrt{2}\sqrt{c_h-b}}\right)\sqrt{b\,(b-c_c)\,c_c}\sqrt{(1-c_h)\,c_h}\,T_h\sqrt{\gamma_c-c_c\gamma_c}\,\gamma_h t_{tot}^2$$

$$+ \sqrt{\pi}\left(\frac{1}{4}e^{\frac{t_{tot}\left(\left(a\gamma_c+\gamma_c+\frac{\gamma_h}{4}\right)b^2-\left(\left(\gamma_c+\frac{\gamma_h}{4}\right)c_c+(a+1)c_h\gamma_c\right)b+c_cc_h\gamma_c\right)}{(a+1)(b-c_c)(b-c_h)}}\,\mathrm{ierf}\left(\frac{\sqrt{\frac{abt_{tot}\gamma_c}{a+1}}}{\sqrt{2}\sqrt{b-c_c}}\right)\sqrt{bc_cc_h\,(c_h-b)}\,t_{tot}^2\,T_c\sqrt{\gamma_c-c_c\gamma_c}\sqrt{(1-c_h)\,\gamma_c\gamma_h}\,a^{3/2}\right.$$

$$-\frac{1}{4}ie^{\frac{t_{tot}\left(\left(a\gamma_c+\gamma_c+\frac{\gamma_h}{4}\right)b^2-\left(\left(\gamma_c+\frac{\gamma_h}{4}\right)c_c+(a+1)c_h\gamma_c\right)b+c_cc_h\gamma_c\right)}{(a+1)(b-c_c)(b-c_h)}}\,\mathrm{erf}\left(\frac{\sqrt{\frac{ac_ct_{tot}\gamma_c}{a+1}}}{\sqrt{2}\sqrt{b-c_c}}\right)\sqrt{bc_cc_h\,(c_h-b)}\,t_{tot}^2\,T_c\sqrt{\gamma_c-c_c\gamma_c}\sqrt{(1-c_h)\,\gamma_c\gamma_h}\,a^{3/2}$$

$$+ t_{tot}\left(\frac{1}{4}e^{t_{tot}\left(\frac{ab\gamma_c}{2(a+1)(b-c_c)}+\gamma_c+\frac{\left(\left(\frac{c_h}{2}-1\right)b+\frac{c_h}{2}\right)\gamma_h}{(a+1)(b-c_h)(c_h-1)}\right)}\,\mathrm{erf}\left(\frac{\sqrt{\frac{t_{tot}\gamma_h}{a+1}}}{\sqrt{2}\sqrt{1-c_h}}\right)\sqrt{b-bc_c}\sqrt{(b-c_c)\,c_cc_h}\sqrt{c_h-b}\,t_{tot}T_h\sqrt{-\gamma_c\gamma_h}\right.$$

$$- t_{tot}\left(\frac{1}{4}e^{\frac{t_{tot}\left(\left(\left(\frac{a}{2}+1\right)(c_h-1)\gamma_c-\frac{c_h\gamma_h}{2}\right)b^2+\left(\left(-c_h\gamma_c+\gamma_c+\frac{c_h\gamma_h}{2}\right)c_c+c_h\left(\left(\frac{a}{2}+1\right)(1-c_h)\gamma_c+\left(c_h-\frac{1}{2}\right)\gamma_h\right)\right)b+c_cc_h\left((c_h-1)\gamma_c-\left(c_h-\frac{1}{2}\right)\gamma_h\right)\right)}{(a+1)(b-c_c)(b-c_h)(c_h-1)}}\right.$$

$$\mathrm{erf}\left(\frac{\sqrt{\frac{c_ht_{tot}\gamma_h}{a+1}}}{\sqrt{2}\sqrt{1-c_h}}\right)\sqrt{b-bc_c}\sqrt{(b-c_c)\,c_cc_h}\sqrt{c_h-b}\,T_h\sqrt{-\gamma_c\gamma_h}-\frac{1}{4}e^{\frac{t_{tot}\left(\frac{\left(c_ca-\frac{a}{2}+c_c-1\right)\gamma_c}{c_c-1}+\frac{ab\gamma_c}{2(b-c_c)}+\frac{b\gamma_h}{2(b-c_h)}\right)}{a+1}}\left(\mathrm{erfi}\left(\frac{\sqrt{at_{tot}}}{\sqrt{2}\sqrt{-\frac{(a+1)(c_c-1)}{\gamma_c}}}\right)\right.$$

$$\left.\left.\left.-\mathrm{erfi}\left(\frac{\sqrt{ac_ct_{tot}}}{\sqrt{2}\sqrt{-\frac{(a+1)(c_c-1)}{\gamma_c}}}\right)\right)\sqrt{ab\,(b-c_c)}\sqrt{(b-c_c)\,(c_h-1)\,c_h}\,T_c\gamma_c\sqrt{-\gamma_h}\right)a\right)\right]\sqrt{2(c_h-1)}$$

$$\Bigg/ \left[\sqrt{2}a(a+1)\sqrt{b-c_c}\sqrt{1-c_h}\,t_{tot}\left(\frac{1}{4}e^{t_{tot}\left(\frac{ab\gamma_c}{2(a+1)(b-c_c)}+\gamma_c+\frac{\left(\left(\frac{c_h}{2}-1\right)b+\frac{c_h}{2}\right)\gamma_h}{(a+1)(b-c_h)(c_h-1)}\right)}\sqrt{bc_c}\sqrt{-\frac{t_{tot}}{a+1}}\sqrt{-\frac{c_ht_{tot}}{a+1}}\right.\right.$$

$$\left.- \frac{1}{4}e^{\frac{t_{tot}\left(\left(\left(\frac{a}{2}+1\right)(c_h-1)\gamma_c-\frac{c_h\gamma_h}{2}\right)b^2+\left(\left(-c_h\gamma_c+\gamma_c+\frac{c_h\gamma_h}{2}\right)c_c+c_h\left(\left(\frac{a}{2}+1\right)(1-c_h)\gamma_c+\left(c_h-\frac{1}{2}\right)\gamma_h\right)\right)b+c_cc_h\left((c_h-1)\gamma_c-\left(c_h-\frac{1}{2}\right)\gamma_h\right)\right)}{(a+1)(b-c_c)(b-c_h)(c_h-1)}}$$

$$\left.\sqrt{c_c}\sqrt{-\frac{bt_{tot}}{a+1}}\sqrt{-\frac{c_ht_{tot}}{a+1}}\right)\sqrt{\gamma_c-c_c\gamma_c}\sqrt{(c_h-b)\,\gamma_h}\right] \tag{A.43}$$

$$C_{h2} = \left[e^{\frac{(b+c_h)t_{tot}\gamma_h}{2(a+1)(b-c_h)}}t_{tot}\left(T_h\left(2\left(e^{\frac{t_{tot}\left(((a+2)b-2c_c)(c_h-1)\gamma_c+(c_c-b)c_h\gamma_h\right)}{2(a+1)(b-c_c)(c_h-1)}}F\left(\frac{\sqrt{-\frac{bt_{tot}\gamma_h}{a+1}}}{\sqrt{2}\sqrt{c_h-b}}\right)-e^{\frac{t_{tot}\left(((3a+2)b-2(a+1)c_c)(c_h-1)\gamma_c+(c_c-b)\gamma_h\right)}{2(a+1)(b-c_c)(c_h-1)}}\right.\right.\right.\right.$$

$$F\left(\frac{\sqrt{-\frac{c_ht_{tot}\gamma_h}{a+1}}}{\sqrt{2}\sqrt{c_h-b}}\right)\right)\sqrt{b\,(b-c_c)\,c_c}\sqrt{-(c_h-1)\,c_h}\sqrt{-(c_c-1)\,\gamma_c}-ie^{\frac{t_{tot}\left(((3a+2)b-2(a+1)c_c)(c_h-1)\gamma_c-(b-c_c)(c_h+1)\gamma_h\right)}{2(a+1)(b-c_c)(c_h-1)}}\sqrt{\pi}$$

$$\left(\mathrm{erf}\left(\frac{\sqrt{\frac{t_{tot}\gamma_h}{a+1}}}{\sqrt{2-2c_h}}\right)-\mathrm{erf}\left(\frac{\sqrt{\frac{c_ht_{tot}\gamma_h}{a+1}}}{\sqrt{2-2c_h}}\right)\right)\sqrt{b-bc_c}\sqrt{(b-c_c)\,c_cc_h}\sqrt{c_h-b}\sqrt{\gamma_c}\right)\gamma_h-i\sqrt{\pi}T_c\left(e^{\frac{t_{tot}\left(2(ab+b-2c_c)(c_h-1)\gamma_c+(c_c-b)c_h\gamma_h\right)}{2(a+1)(b-c_c)(c_h-1)}}\right.$$

$$\sqrt{abc_cc_h\,(c_h-b)}\sqrt{-(c_c-1)\,\gamma_c}\sqrt{-(c_h-1)\,\gamma_c\gamma_h}\left(\mathrm{erf}\left(\frac{\sqrt{\frac{abt_{tot}\gamma_c}{a+1}}}{\sqrt{2}\sqrt{b-c_c}}\right)-\mathrm{erf}\left(\frac{\sqrt{\frac{ac_ct_{tot}\gamma_c}{a+1}}}{\sqrt{2}\sqrt{b-c_c}}\right)\right)$$

$$+ e^{\frac{t_{tot}\left((b-c_c)(c_c-1)c_h\gamma_h-\left(-2(a+1)c_c^2+(3ba+a+2b+2)c_c-2(a+1)b\right)(c_h-1)\gamma_c\right)}{2(a+1)(c_c-1)(c_c-b)(c_h-1)}}\left(\mathrm{erfi}\left(\frac{\sqrt{at_{tot}}}{\sqrt{2}\sqrt{-\frac{(a+1)(c_c-1)}{\gamma_c}}}\right)-\mathrm{erfi}\left(\frac{\sqrt{ac_ct_{tot}}}{\sqrt{2}\sqrt{-\frac{(a+1)(c_c-1)}{\gamma_c}}}\right)\right)$$

$$\left.\left.\sqrt{ab\,(b-c_c)\,c_c\gamma_c}\sqrt{-(b-c_h)\,(c_h-1)\,c_h\gamma_h}\right)\right)\sqrt{2(b-c_h)}$$

$$\Bigg/\left[\sqrt{2}\sqrt{(a+1)\,(b-c_c)}\sqrt{1-c_h}\sqrt{-c_h}\,t_{tot}\left(e^{\frac{t_{tot}\left(((a+2)b-2c_c)(b-c_h)(c_h-1)\gamma_c-(b-c_c)(b-2c_h+1)c_h\gamma_h\right)}{2(a+1)(b-c_c)(b-c_h)(c_h-1)}}\sqrt{c_c}\sqrt{-\frac{bt_{tot}}{a+1}}\right.\right.$$

$$\left.\left.-e^{t_{tot}\left(\left(\frac{ab}{2(a+1)(b-c_c)}+1\right)\gamma_c+\frac{((b+1)c_h-2b)\gamma_h}{2(a+1)(b-c_h)(c_h-1)}\right)}\sqrt{bc_c}\sqrt{-\frac{t_{tot}}{a+1}}\right)\sqrt{-(c_c-1)\,\gamma_c}\sqrt{-(b-c_h)\,\gamma_h}\right] \tag{A.44}$$

$$C_{c1} = \left[a\sqrt{\pi} \left(c_h \sqrt{-(c_c-1)\,\gamma_c} \left((a+1)e^{\frac{t_{tot}((a+2)b-2c_c)(b-c_h)(c_h-1)\gamma_c-(b-c_c)(b-2c_h+1)c_h\gamma_h}{2(a+1)(b-c_c)(b-c_h)(c_h-1)}} \mathrm{erf}\left(\frac{\sqrt{\frac{ac_ct_{tot}\gamma_c}{a+1}}}{\sqrt{2}\sqrt{b-c_c}} \right) \sqrt{c_c}\sqrt{\frac{bt_{tot}}{a+1}} \right. \right.$$

$$\sqrt{(1-c_h)\,\gamma_c\,(b-c_h)\,\gamma_h} - ie^{t_{tot}\left(\left(\frac{ab}{2(a+1)(b-c_c)}+1\right)\gamma_c+\frac{((b+1)c_h-2b)\gamma_h}{2(a+1)(b-c_h)(c_h-1)}\right)} \mathrm{erf}\left(\frac{\sqrt{\frac{abt_{tot}\gamma_c}{a+1}}}{\sqrt{2}\sqrt{b-c_c}} \right) \sqrt{-bc_c\,(b-c_h)t_{tot}}$$

$$\sqrt{-\frac{(a+1)(c_h-1)\,\gamma_c\gamma_h}{t_{tot}}} \right) + \sqrt{-\frac{c_h t_{tot}}{a+1}}(a+1)e^{\frac{t_{tot}\left(\frac{-\frac{a}{2}+(a+1)c_c-1}{c_c-1}\gamma_c-\frac{(b-2c_h+1)c_h\gamma_h}{2(b-c_h)(c_h-1)}\right)}{a+1}}$$

$$i\left(\mathrm{erfi}\left(\frac{\sqrt{at_{tot}}}{\sqrt{2}\sqrt{-\frac{(a+1)(c_c-1)}{\gamma_c}}} \right) - \mathrm{erfi}\left(\frac{\sqrt{ac_ct_{tot}}}{\sqrt{2}\sqrt{-\frac{(a+1)(c_c-1)}{\gamma_c}}} \right) \right) \sqrt{b\,(b-c_c)\,c_c\,(b-c_h)\,(1-c_h)\,c_h\gamma_h\gamma_c} \right) T_c$$

$$+ i\sqrt{-(a+1)c_h t_{tot}}\,T_h \left(-2e^{t_{tot}\left(\gamma_c+\frac{((b+1)c_h-2b)\gamma_h}{2(a+1)(b-c_h)(c_h-1)}\right)} \sqrt{ab\,(b-c_c)\,c_c\,(c_h-1)\,c_h\,(c_c-1)\,\gamma_c}\,F\left(\frac{\sqrt{-\frac{bt_{tot}\gamma_h}{a+1}}}{\sqrt{2}\sqrt{c_h-b}} \right) \right.$$

$$+ e^{\frac{t_{tot}\left((-2c_h^2+bc_h+b)\gamma_h-2(a+1)(b-c_h)(c_h-1)\gamma_c\right)}{2(a+1)(c_h-1)(c_h-b)}} \sqrt{\pi}\left(\mathrm{erf}\left(\frac{\sqrt{\frac{t_{tot}\gamma_h}{a+1}}}{\sqrt{2-2c_h}} \right) - \mathrm{erf}\left(\frac{\sqrt{\frac{c_h t_{tot}\gamma_h}{a+1}}}{\sqrt{2-2c_h}} \right) \right) \sqrt{ab\,(c_c-1)\,(b-c_c)\,c_c c_h(c_h-b)\gamma_c}$$

$$+ 2e^{t_{tot}\left(\gamma_c+\frac{(c_h^2-b)\gamma_h}{2(a+1)(b-c_h)(c_h-1)}\right)} F\left(\frac{\sqrt{-\frac{c_h t_{tot}\gamma_h}{a+1}}}{\sqrt{2}\sqrt{c_h-b}} \right) \sqrt{ab\,(b-c_c)\,c_c\,(c_h-1)\,c_h\,(c_c-1)\,\gamma_c}\,\gamma_h \right) \sqrt{(b-c_c)}$$

$$\left/ \left[a(a+1)\sqrt{b-c_c}\sqrt{1-c_h}c_h \left(e^{\frac{t_{tot}((a+2)b-2c_c)(b-c_h)(c_h-1)\gamma_c-(b-c_c)(b-2c_h+1)c_h\gamma_h}{2(a+1)(b-c_c)(b-c_h)(c_h-1)}} \sqrt{c_c}\sqrt{-\frac{bt_{tot}}{a+1}} \right. \right. \right.$$

$$\left. \left. -e^{t_{tot}\left(\left(\frac{ab}{2(a+1)(b-c_c)}+1\right)\gamma_c+\frac{((b+1)c_h-2b)\gamma_h}{2(a+1)(b-c_h)(c_h-1)}\right)} \sqrt{bc_c}\sqrt{-\frac{t_{tot}}{a+1}} \right) \sqrt{-(c_c-1)\,\gamma_c}\sqrt{-(b-c_h)\,\gamma_h} \right] \tag{A.45}$$

$$C_{c2} = \left[e^{-\frac{\left(-\frac{a}{2}+(a+1)c_c-1\right)t_{tot}\gamma_c}{(a+1)(c_c-1)}} \left(a\sqrt{\pi}\left(e^{\frac{t_{tot}(2(2(a+1)b-(a+2)c_c)(b-c_h)(c_h-1)\gamma_c+(b-c_c)((b+1)c_h-2b)\gamma_h)}{2(a+1)(b-c_c)(b-c_h)(c_h-1)}} \sqrt{1-c_c}\,\mathrm{erf}\left(\frac{\sqrt{\frac{abt_{tot}\gamma_c}{a+1}}}{\sqrt{2}\sqrt{b-c_c}} \right) \right. \right. \right.$$

$$-e^{\frac{t_{tot}(2(2(a+1)b-(a+2)c_c)(b-c_h)(c_h-1)\gamma_c+(b-c_c)((b+1)c_h-2b)\gamma_h)}{2(a+1)(b-c_c)(b-c_h)(c_h-1)}}\,\mathrm{erf}\left(\frac{\sqrt{\frac{ac_ct_{tot}\gamma_c}{a+1}}}{\sqrt{2}\sqrt{b-c_c}} \right) \sqrt{1-c_c}$$

$$+ \left(e^{\frac{t_{tot}}{2(a+1)}\left(\frac{\left(2(a+2)c_c^2-(3ba+a+4b+4)c_c+2(a+2)b\right)\gamma_c}{(c_c-1)(c_c-b)}-\frac{(b-2c_h+1)c_h\gamma_h}{(b-c_h)(c_h-1)}\right)} \mathrm{erfi}\left(\frac{\sqrt{\frac{at_{tot}\gamma_c}{a+1}}}{\sqrt{2-2c_c}} \right) \right.$$

$$\left. -e^{\frac{t_{tot}}{2(a+1)}\left(\frac{\gamma_c(c_c(5ab-4(a+1)c_c+3a+4b+4)-4(a+1)b)}{(c_c-1)(b-c_c)}+\frac{\gamma_h((b+1)c_h-2b)}{(c_h-1)(b-c_h)}\right)} \mathrm{erfi}\left(\frac{\sqrt{\frac{ac_ct_{tot}\gamma_c}{a+1}}}{\sqrt{2-2c_c}} \right) \right) \sqrt{b-c_c} \right) \sqrt{(b-c_h)\,(1-c_h)\,bc_c c_h\gamma_h}\,T_c\gamma_c$$

$$+ \sqrt{b-c_c}\,T_h \left(-2e^{\frac{t_{tot}\left(\frac{((3a+4)b-2(a+2)c_c)\gamma_c}{b-c_c}+\frac{((b+1)c_h-2b)\gamma_h}{(b-c_h)(c_h-1)}\right)}{2(a+1)}} \sqrt{-abc_c\,(c_h-1)c_h}\sqrt{-(c_c-1)\,\gamma_c}\,F\left(\frac{\sqrt{-\frac{bt_{tot}\gamma_h}{a+1}}}{\sqrt{2}\sqrt{c_h-b}} \right) \right.$$

$$+ e^{\frac{t_{tot}\left(((3a+4)b-2(a+2)c_c)(b-c_h)(c_h-1)\gamma_c-(b-c_c)\left(-2c_h^2+bc_h+b\right)\gamma_h\right)}{2(a+1)(b-c_c)(b-c_h)(c_h-1)}} \left(\mathrm{erf}\left(\frac{\sqrt{\frac{t_{tot}\gamma_h}{a+1}}}{\sqrt{2-2c_h}} \right) - \mathrm{erf}\left(\frac{\sqrt{\frac{c_h t_{tot}\gamma_h}{a+1}}}{\sqrt{2-2c_h}} \right) \right)$$

$$\sqrt{\pi(1-c_c)(b-c_h)abc_c c_h\gamma_c} + 2e^{\frac{t_{tot}\left(\frac{((3a+4)b-2(a+2)c_c)\gamma_c}{b-c_c}+\frac{(c_h^2-b)\gamma_h}{(b-c_h)(c_h-1)}\right)}{2(a+1)}} F\left(\frac{\sqrt{-\frac{c_h t_{tot}\gamma_h}{a+1}}}{\sqrt{2}\sqrt{c_h-b}} \right) \sqrt{1-c_c}\sqrt{-abc_c\,(c_h-1)\,c_h\gamma_c} \right) \gamma_h \right) \right]$$

$$\sqrt{c_c-1}\left/ \left[a\left(-e^{\frac{t_{tot}(((a+2)b-2c_c)(b-c_h)(c_h-1)\gamma_c-(b-c_c)(b-2c_h+1)c_h\gamma_h)}{2(a+1)(b-c_c)(b-c_h)(c_h-1)}} + e^{t_{tot}\left(\left(\frac{ab}{2(a+1)(b-c_c)}+1\right)\gamma_c+\frac{((b+1)c_h-2b)\gamma_h}{2(a+1)(b-c_h)(c_h-1)}\right)} \right) \right. \right.$$

$$\left. \sqrt{b-c_c}\sqrt{(b-c_h)\,(c_h-1)}\sqrt{-bc_c c_h}\sqrt{-(c_c-1)\,\gamma_c\gamma_h} \right] \tag{A.46}$$

A.6   Feynman-Kac Formula

The Feynman-Kac formula [70] establishes a link between integrals over stochastic processes and parabolic partial differential equations. Let us consider a stochastic process that is described by the Ito stochastic differential equation

$$dx^i = u^i(x,t)dt + \beta^{ij}(x,t) \cdot dW_j, \tag{A.47}$$

with drift and diffusion coefficients

$$D_i^{(1)} = u^i(x,t), \quad D_{ij}^{(2)} = \beta^{ik}(x,t)\beta^{jk}(x,t). \tag{A.48}$$

We are interested in the properties of the stochastic integral

$$J(x,t|x',t') = \int_{t_0}^{t} h(x,\tau)d\tau + g_i(x,\tau) \cdot dx^i \tag{A.49}$$

over trajectories that start at $x(t_0) = x_0$ and end in $x(t) = x$. The generating function of J can be written as a discretized path integral

$$G_s(x,t|x',t') = \langle e^{sJ}\delta(x(t) - x) \rangle = \lim_{N\to\infty} G_s^{(N)}(x,t|x',t'), \tag{A.50}$$

with

$$G_s^{(N)}(x,t|x_0,t_0) = \int \left(\prod_{k=1}^{N-1} dx_k \right) \exp\left\{ -s\sum_{k=0}^{N-1}[h(x_k,\tau_k)\Delta + g_i(x_k,\tau_k)(x_{k+1}^i - x_k^i)] \right\}$$
$$\prod_{k=1}^{N-1} p(x_{k+1},\tau_{k+1}|x_k,\tau_k), \tag{A.51}$$

where the discretization follows

$$\tau_k = t' + k\Delta, \quad \Delta = (t'-t)/N. \tag{A.52}$$

To derive the forward Feynman-Kac formula we notice the recursive relation

$$G_s^{(N)}(x,t|x_0,t_0) = \int dx_{N-1} p(x_N,\tau_N|x_{N-1}) \exp\{-s[h(x_{N-1},\tau_{N-1})\Delta + g_i(x_{N-1},\tau_{N-1})(x_N^i - x_{N-1}^i)]\}$$
$$G^{(N-1)}(x_{N-1},\tau_{N-1}|x_0,\tau_0). \tag{A.53}$$

We multiply both sides by an arbitrary function $\theta(x_N)$, integrate over x_N and expand the exponential and $\theta(x_N)$ around x_{N-1} and τ_{N-1}. Up to first order in Δ we obtain

$$\int dx_N \theta(x_N) G_s^{(N)}(x,t|x_0,t_0) = \int dx_N dx_{N-1} p(x_N,\tau_N|x_{N-1}) G^{(N-1)}(x_{N-1},\tau_{N-1}|x_0,\tau_0) \times I_\theta \times I_{exp}$$

$$I_\theta = \theta(x_{N-1}) + (x_N^i - x_{N-1}^i)\frac{\partial\theta}{\partial x^i}(x_{N-1}) + \frac{1}{2}(x_N^i - x_{N-1}^i)(x_N^j - x_{N-1}^j)\frac{\partial^2\theta}{\partial x^i \partial x^j}(x_{N-1})$$

$$I_{exp} = 1 - sh(x_{N-1},\tau_{N-1})\Delta - s(x_N^i - x_{N-1}^i)g_i(x_{N-1},\tau_{N-1})$$
$$+ \frac{s^2}{2}(x_N^i - x_{N-1}^i)(x_N^j - x_{N-1}^j)g_i(x_{N-1},\tau_{N-1})g_j(x_{N-1},\tau_{N-1}). \tag{A.54}$$

After integrating over x_N we can replace the first and second moments by the drift and diffusion coefficients from equation (A.48), $\langle (x_N^i - x_{N-1}^i) \rangle = D_i^{(1)}(x_{N-1}^i)\Delta$ and $\langle (x_N^i - x_{N-1}^i)(x_N^j - x_{N-1}^j) \rangle = D_{ij}^{(2)}(x_{N-1}^i)\Delta$ and we obtain

$$
\int dx_N \theta(x_N) G_s^{(N)}(x,t|x_0,t_0) = \int dx_{N-1} G^{(N-1)}(x_{N-1},\tau_{N-1}|x_0,\tau_0) \left[\theta(x_{N-1}) + D_i^{(1)} \frac{\partial \theta}{\partial x^i}(x_{N-1})\Delta \right.
$$
$$
+ \frac{1}{2} D_{ij}^{(2)} \frac{\partial^2 \theta}{\partial x^i \partial x^j}(x_{N-1})\Delta - sh(x_{N-1},\tau_{N-1})\theta(x_{N-1})\Delta - sD_i^{(1)} g_i(x_{N-1},\tau_{N-1})\theta(x_{N-1})\Delta
$$
$$
\left. + \frac{s^2}{2} D_{ij}^{(2)} g_i(x_{N-1},\tau_{N-1}) g_j(x_{N-1},\tau_{N-1})\theta(x_{N-1})\Delta - sD_{ij}^{(2)} g_i(x_{N-1},\tau_{N-1}) \frac{\partial \theta}{\partial x^j}(x_{N-1})\Delta \right]. \quad (A.55)
$$

Integrating by parts in x_{N-1}, expanding the left hand side around τ_{N-1}, comparing the integrands and taking the limit $N \to \infty$ one finally obtains the forward Feynman-Kac equation for $G_s(x,t|x',t')$

$$
\frac{\partial G_s}{\partial t} - L^\dagger G_s = - \left(sh + sD_i^{(1)} g_i - \frac{s^2}{2} D_{ij}^{(2)} g_i g_j \right) G_s + s \frac{\partial}{\partial x^j} \left(D_{ij}^{(2)} g_i G_s \right), \quad (A.56)
$$

where $h, g, D^{(1)}$ and $D^{(2)}$ are evaluated at final configurations x, t.

Bibliography

[1] J. B. J. Fourier, *Théorie analytique de la chaleur.* Chez Firmin Didot, 1822.

[2] S. Carnot, *Réflexions sur la puissance motrice du feu et sur les machines propres à développer cette puissance.* Gauthier-Villars, 1824.

[3] W. Coffey, Y. Kalmykov, and J. Waldron, *The Langevin Equation.* World Scientific, 2004.

[4] A. Einstein, "On the Movement of Small Particles Suspended in Stationary Liquids Required by the Molecular-Kinetic Theory of Hear," *Ann. Phys.* **17** (1905) 549–560.

[5] M. R. Smoluchowski, "The kinetic theory of Brownian molecular motion and suspensions," *Ann. Phys.* **21** (1906) 756.

[6] C. Gardiner, *Handbook of Stochastic Methods.* Springer, 3rd ed., 2004.

[7] J. S. Townsend, "Jean Baptiste Perrin. 1870 - 1942," *Biogr. Mem. Fellows R. Soc.* **4** (1943) 301–326.

[8] P. Langevin, "Sur la théorie du mouvement Brownien," *Comptes Rendues* **146** (1908) 530.

[9] K. Ito, "On the ergodicity of certain stationary processes," *Proc. Imp. Acad. Tokyo* **20** (1944) 54–55.

[10] K. Sekimoto, "Kinetic Characterization of Heat Bath and the Energetics of Thermal Ratchet Models," *J. Phys. Soc. Jpn.* **66** (1997) 1234.

[11] K. Sekimoto, "Langevin Equation and Thermodynamics," *Prog. Theor. Phys. Suppl.* **130** (1998) 17–27.

[12] K. Sekimoto, *Stochastic Energetics.* Springer, 2010.

[13] T. Schmiedl and U. Seifert, "Optimal Finite-Time Processes In Stochastic Thermodynamics," *Phys. Rev. Lett.* **98** (2007) 108301.

[14] P. R. Zulkowski, D. A. Sivak, and M. R. DeWeese, "Optimal control of transitions between nonequilibrium steady states," *PloS ONE* **8** (2013) e82754.

[15] H. Then and A. Engel, "Computing the optimal protocol for finite-time processes in stochastic thermodynamics," *Phys. Rev. E* **77** (2008) 041105.

[16] E. Aurell, C. Mejia-Monasterio, and P. Muratore-Ginanneschi, "Optimal protocols and optimal transport in stochastic thermodynamics," *Phys. Rev. Lett.* **106** (2011) 250601.

[17] T. Zhan-Chun, "Recent advance on the efficiency at maximum power of heat engines," *Chin. Phys. B* **21** (2012) 020513.

[18] M. Esposito and K. Lindenberg, "Universality of Efficiency at Maximum Power," *Phys. Rev. Lett.* **102** (2009) 130602.

[19] Y. Apertet, H. Ouerdane, C. Goupil, and P. Lecoeur, "Efficiency at maximum power of thermally coupled heat engines," *Phys. Rev. E* **85** (2012) 041144.

[20] M. Esposito, R. Kawai, K. Lindenberg, and C. Van den Broeck, "Efficiency at Maximum Power of Low-Dissipation Carnot Engines," *Phys. Rev. Lett.* **105** (2010) 150603.

[21] U. Seifert, "Stochastic thermodynamics, fluctuation theorems and molecular machines," *Rep. Prog. Phys.* **75** no. 12, (2012) 126001.

[22] T. Schmiedl and U. Seifert, "Efficiency at maximum power: An analytically solvable model for stochastic heat engines," *Europhys. Lett.* **81** (2008) .

[23] V. Blickle and C. Bechinger, "Realization of a micrometer-sized stochastic heat engines," *Nat. Phys.* **8** (2012) 143–146.

[24] A. Celani, S. Bo, R. Eichhorn, and E. Aurell, "Anomalous Thermodynamics at the Microscale," *Phys. Rev. Lett.* **109** (2012) 260603.

[25] H. Risken, *The Fokker-Planck Equation*. Springer, 2nd ed., 1996.

[26] D. Ryter and U. Deker, "Properties of the noise-induced spurious drift. I.," *J. Math. Phys.* **21** no. 11, (1980) 2662.

[27] U. Deker and D. Ryter, "Properties of the noise-induced spurious drift. II. Simplifications of Langevin equations," *J. Math. Phys.* **21** no. 11, (1980) 2666.

[28] A. D. Fokker, "Die mittlere Energie rotierender elektrischer Dipole im Strahlungsfeld," *Ann. Phys.* **43** (1914) 810.

[29] M. Planck, "An essay on statistical dynamics and its amplification in the quantum theory.," *SKPrAW* (1917) 324.

[30] H. A. Kramers, "Brownian motion in a field of force and the difusion model of chemical reactions," *Physica* **7** no. 4, (1940) 284.

[31] H. J. Ilgauds, *Norbert Wiener.* No. 45 in Biographien hervorragender Naturwissenschaftler, Techniker und Mediziner. Teubner, 1980.

[32] N. G. van Kampen, "Ito Versus Stratonovich," *J. Stat. Phys.* **24** (1980) 175–187.

[33] G. Wilemski, "On the derivation of Smoluchowski equations with corrections in the classical theory of Brownian motion," *J. Stat. Phys.* **14** (1976) 153–169.

[34] L. Bocquet, "High Friction Limit of the Kramers Equation," *arXiv:cond-mat/9605186* (1996) .

[35] U. Seifert, "Entropy Production along a Stochastic Trajectory and an Integral Fluctuation Theorem," *Phys. Rev. Lett.* **95** (2005) 040602.

[36] N. Kiesel, F. Blaser, U. Delić, D. Grass, R. Kaltenbaek, and M. Aspelmeyer, "Cavity cooling of an optically levitated submicron particle," *Proc. Natl. Acad. Sci. USA* **110** no. 35, (2013) 14180–14185.

[37] I. Favero, S. Stapfner, D. Hunger, P. Paulitschke, J. Reichel, H. Lorenz, E. Weig, and K. Karrai, "Fluctuating nanomechanical system in a high finesse optical microcavity," *Opt. Express* **17** (2009) 12813.

[38] G. Anetsberger, O. Arcizet, Q. P. Unterreithmeier, R. Riviere, A. Schliesser, E. M. Weig, J. P. Kotthaus, and T. J. Kippenberg, "Near-field cavity optomechanics with nanomechanical oscillators," *Nat. Phys.* **5** (2011) 909–914.

[39] J. D. Thompson, B. M. Zwickl, A. M. Jayich, F. Marquardt, S. M. Girvin, and J. G. E. Harris, "Strong dispersive coupling of a high-finesse cavity to a micromechanical membrane," *Nature* **452** (2008) 06715.

[40] V. Blickle, T. Speck, L. Helden, U. Seifert, and C. Bechinger, "Thermodynamics of a Colloidal Particle in a Time-Dependent Nonharmonic Potential," *Phys. Rev. Lett.* **96** (2006) 070603.

[41] H. Spohn and J. L. Lebowitz, "Irreversible Thermodynamics for Quantum Systems Weakly Coupled to Thermal Reservoirs," *Adv. Chem. Phys.* **38** (1978) 109.

[42] D. J. Evans, E. G. D. Cohen, and G. P. Morriss, "Probability of second law violations in shearing steady states," *Phys. Rev. Lett.* **71** (1993) 2401–2404.

[43] D. J. Evans and D. J. Searles, "Equilibrium microstates which generate second law violating steady states," *Phys. Rev. E* **50** (1994) 1645–1648.

[44] A. H. Carter, *Classical and Statistical Thermodynamics*. Pearson Springer-Wesley, 2000.

[45] F. L. Curzon and B. Ahlborn, "Efficiency of a Carnot engine at maximum power output," *Am. J. Phys.* **43** (1975) 22.

[46] P. Reimann, "Brownian motors: noisy transport far from equilibrium," *Phys. Rep.* **361** (2002) 57–265.

[47] A. Ashkin, *Optical trapping and manipulation of neutral particles using lasers*. World Scientific, 2007.

[48] G. E. P. Box and M. E. Muller, "A Note on the Generation of Random Normal Deviates," *Ann. Math. Statist.* **29** (1958) 610–611.

[49] S. Bo and A. Celani, "Entropic anomaly and maximal efficiency of microscopic heat engines," *Phys. Rev. E* **87** (May, 2013) 050102.

[50] M. Aspelmeyer, T. Kippenberg, and F. Marquardt, "Cavity Optomechanics," *arXiv:1303.0733v1* (2013) .

[51] P. Kunkel and V. Mehrmann, *Differential-Algebraic Equations, Analysis and Numerical Solution*. European Mathematical Society, 2006.

[52] R. Stratonovich, *Topics in the Theory of Random Noise Volume I*. Gordon and Breach, 1963.

[53] R. K. Pathria, *Statistical Mechanics*. Butterworth Heinemann, 2nd ed., 1996.

[54] G. Andrews, R. Askey, and R. Roy, *Special Functions, Encyclopedia of Mathematics and its Applications*. Cambridge University Press, 1999.

[55] R. van Zon and E. G. D. Cohen, "Non-equilibrium thermodynamics and fluctuation," *Physica A* **340** (2004) 66–75.

[56] R. van Zon and E. G. D. Cohen, "Extension of the Fluctuation Theorem," *Phys. Rev. Lett.* **91** (2003) .

[57] R. van Zon and E. G. D. Cohen, "Stationary and transient work-fluctuation theorems for a dragged Brownian particle," *Phys. Rev. E* **67** (2003) .

[58] J. M. R. Parrondo, C. van den Broeck, and R. Kawai, "Entropy production and the arrow of time," *New J. Phys.* **11** (2009) 073008.

[59] H. G. Schuster, *Nonequilibrium Statistical Physics of Small Systems*. Wiley, 2013.

[60] R. Chetrite and K. Gawedzki, "Fluctuation relations for diffusion processes," *Commun. Math. Phys.* **282** (2008) 469518.

[61] R. Courant and D. Hilbert, *Methods of Mathematica Physics, Volume 1*. Wiley, 1953.

[62] P. Turq, F. Lantelme, and H. L. Friedman, "Brownian dynamics: Its application to ionic solutions," *J. Chem. Phys.* **66** (1977) 3039.

[63] G. Bussi and M. Parrinello, "Accurate sampling using Langevin dynamics," *Phys. Rev. E* **75** (2007) 056707.

[64] D. A. Sivak, J. D. Chodera, and G. E. Crooks, "Using Nonequilibrium Fluctuation Theorems to Understand and Correct Errors in Equilibrium and Nonequilibrium Simulations of Discrete Langevin Dynamics," *arXiv:1107.2967v5* (2013) .

[65] M. E. Tuckerman, G. J. Martyna, and B. J. Berne, "Molecular dynamics algorithm for condensed systems with multiple time scales," *J. Chem. Phys.* **93** (1990) 1287.

[66] H. Trotter, "On the product of semi-groups of operators," *Proc. Natl. Acad. Sci. USA* **10** (1959) 545–551.

[67] J. M. Sanz-Serna, "Symplectic integrators for Hamiltonian problems: an overview," *Acta Numer.* **1** (1992) 243–286.

[68] H. Yoshida, "Construction of higher order symplectic integrators," *Phys. Lett. A* **150** (1990) 262–268.

[69] A. Gomez-Marin, T. Schmiedl, and U. Seifert, "Optimal protocols for minimal work processes in underdamped stochastic thermodynamics," *J. Chem. Phys.* **129** (2008) 024114.

[70] M. Kac, "On distributions of certain Wiener functionals," *Trans. Amer. Math. Soc.* **65** (1949) 1–13.

Printed in the United States
By Bookmasters

Printed in the United States
By Bookmasters